AF690612

RENEWABLE ENERGY LAW

With the rapid growth of the renewable energy sector, it has become increasingly important to understand how renewable energy is defined in national laws around the world and what regulatory mechanisms these countries are deploying to achieve their renewable energy goals. In *Renewable Energy Law: An International Assessment*, Dr Penelope Crossley compares the national renewable energy laws for each of the 113 countries that have such a law, shedding light on the question of whether energy laws are converging globally to facilitate trade or engaging in regulatory competition. The book includes over sixty extracts from different national laws, case studies on the European Union and the Chinese wind sector, and many examples of the particular challenges facing specific countries. This work should be read by scholars, policymakers, regulators, employees of commercial entities operating in the energy sector, and anyone else interested in the legal and regulatory landscape of renewable energy.

Penelope Crossley is a Director of the Energy Users Association of Australia and a member of the Technical Working Group of the Energy Security Board. Prior to entering academia, Penelope practised as a Solicitor in London and Beijing specialising in Global Energy and Infrastructure Law.

Renewable Energy Law

AN INTERNATIONAL ASSESSMENT

PENELOPE CROSSLEY
The University of Sydney Law School

CAMBRIDGE
UNIVERSITY PRESS

CAMBRIDGE
UNIVERSITY PRESS

University Printing House, Cambridge CB2 8BS, United Kingdom

One Liberty Plaza, 20th Floor, New York, NY 10006, USA

477 Williamstown Road, Port Melbourne, VIC 3207, Australia

314–321, 3rd Floor, Plot 3, Splendor Forum, Jasola District Centre, New Delhi – 110025, India

79 Anson Road, #06–04/06, Singapore 079906

Cambridge University Press is part of the University of Cambridge.

It furthers the University's mission by disseminating knowledge in the pursuit of education, learning, and research at the highest international levels of excellence.

www.cambridge.org
Information on this title: www.cambridge.org/9781107185760
DOI: 10.1017/9781316888490

© Penelope Crossley 2019

This publication is in copyright. Subject to statutory exception and to the provisions of relevant collective licensing agreements, no reproduction of any part may take place without the written permission of Cambridge University Press.

First published 2019

A catalogue record for this publication is available from the British Library.

Library of Congress Cataloging-in-Publication Data
NAMES: Crossley, Penelope, author.
TITLE: Renewable energy law : an international assessment / Penelope Crossley, the University of Sydney Law School.
DESCRIPTION: Cambridge, United Kingdom ; New York, NY : Cambridge University Press, 2019. | Includes bibliographical references and index.
IDENTIFIERS: LCCN 2019004944 | ISBN 9781107185760 (alk. paper)
SUBJECTS: LCSH: Renewable energy sources – Law and legislation. | Renewable energy sources – Economic aspects. | Renewable energy sources – Government policy. | Energy industries – Law and legislation.
CLASSIFICATION: LCC K3981. C76 2019 | DDC 346.04/6794–dc23
LC record available at https://lccn.loc.gov/2019004944

ISBN 978-1-107-18576-0 Hardback

Cambridge University Press has no responsibility for the persistence or accuracy of URLs for external or third-party internet websites referred to in this publication and does not guarantee that any content on such websites is, or will remain, accurate or appropriate.

This book is dedicated to Hayden and Madeleine.

Contents

List of Tables — *page* x
Acknowledgements — xii
List of Acronyms and Abbreviations — xiii
List of National Renewable Energy Legislation — xvi
List of Other Legislation and Treaties — xxx
Units of Measurement — xxxii

1 **Introduction** — 1
- 1.1 The Problem and Significance of the Research — 4
- 1.2 Hypothesis and Research Methodology — 7
- 1.3 Structure of the Book — 11

PART I WHAT IS RENEWABLE ENERGY? A CASE OF CONCEPTUAL CONSENSUS — 17

2 **The Renewable Energy Sources Used for Electricity Generation** — 19
- 2.1 Wind Energy — 20
- 2.2 Solar Energy — 24
- 2.3 Biomass — 27
- 2.4 Landfill Gas, Sewage Treatment Gas and Biogas — 34
- 2.5 Hydropower — 36
- 2.6 Geothermal Energy — 43
- 2.7 Ocean and Riverine Energies — 47
- 2.8 Hydrogen/Fuel Cells — 53
- 2.9 Peat — 55
- 2.10 Nuclear Energy — 56

2.11 Any Other Source Prescribed by the Regulations or Otherwise Permitted 60
2.12 Conclusion 60

PART II WHY DO COUNTRIES INTERVENE IN THE RENEWABLE ENERGY SECTOR? A CASE OF NORMATIVE DIVERGENCE 63

3 **The Economic Justification for Regulating Renewable Energy** 65
3.1 The Characteristics of Electricity That Warrant Special Regulatory Treatment 65
3.2 Market Failures Affecting the Renewable Energy Sector 73
3.3 Market Barriers Within the Renewable Energy Sector 79
3.4 Is Regulatory Intervention in the Renewable Energy Sector Warranted from an Economic Perspective? 93
3.5 Conclusion 96

4 **Why Do Countries Legislate to Accelerate the Deployment of Renewable Energy?** 98
4.1 What Does It Mean to Legislate? 98
4.2 The Identification and Role of Legislative Objectives 100
4.3 The Legislative Objectives in Renewable Energy Law 103
4.4 A Comprehensive Study of the Legislative Objectives in Renewable Energy Laws 107
4.5 Multiple and Competing Legislative Objectives 159
4.6 Conclusion 161

PART III WHAT ROLE DO REGULATORY SUPPORT MECHANISMS PLAY IN NATIONAL RENEWABLE ENERGY LAWS? A CASE OF SUBSTANTIVE DIVERGENCE 165

5 **How Do Countries Regulate to Support Renewable Energy?** 167
5.1 The Selection of Regulatory Support Mechanisms 167
5.2 Types of Regulatory Support Mechanisms Used in the Renewable Energy Sector 176
5.3 Evaluating the Success of Regulatory Support Mechanisms 219
5.4 Conclusion 222

6 **The Future Development of Regulatory Support Mechanisms – Unification, Harmonisation, Convergence, Divergence or Regulatory Competition?** 224
6.1 What Are the Advantages and Disadvantages of National Renewable Energy Laws Becoming More Similar Internationally? 225
6.2 The Future Development of Regulatory Support Mechanisms: Same, Same or Different? 229
6.3 Unification 230
6.4 Harmonisation 231
6.5 Convergence 237
6.6 Divergence 240
6.7 Regulatory Competition 241
6.8 Conclusion 248

7 **Conclusion** 250
7.1 What Is Renewable Energy? A Case of Conceptual Consensus 253
7.2 Why Intervene in the Renewable Energy Sector? A Case of Normative Divergence 255
7.3 What Role Do Regulatory Support Mechanisms Play in National Renewable Energy Laws? A Case of Substantive Divergence 260
7.4 Conclusion 264

Index 265

Tables

2.1 Classification of Hydropower Projects by Size *page* 37
4.1 Comparison of the Legislative Objectives and Their Priorities of the Countries Whose Name Begins with the Letter 'G' 104
4.2 Legislative Objectives by the Number of Countries Adopting Them and Weighted Rank 109
4.3 Countries Citing 'Energy Security' in Their Legislative Objectives 113
4.4 Countries Citing 'Diversify Supply' in Their Legislative Objectives 115
4.5 Countries Citing 'Reduce Fossil Fuel Imports or Nuclear Imports' in Their Legislative Objectives 117
4.6 Countries Citing 'Encourage Greater Use of Indigenous Energy Sources' in Their Legislative Objectives 118
4.7 Countries Citing 'Improve Energy System Safety and Reliability' in Their Legislative Objectives 119
4.8 Countries Citing 'Improve the Structure of the Energy Sector' in Their Legislative Objectives 121
4.9 Countries Citing 'More Efficient Use of Natural Resources and Energy Conservation' in Their Legislative Objectives 123
4.10 Countries Citing 'Sustainable Development' in Their Legislative Objectives 124
4.11 Countries Citing 'Competition and Consumer Issues' in Their Legislative Objectives 127
4.12 Countries Citing 'Promote Private Investment' in Their Legislative Objectives 128
4.13 Countries Citing 'Strengthen the Economy' in Their Legislative Objectives 129
4.14 Countries Citing 'National Development' in Their Legislative Objectives 130

4.15 Countries Citing 'Increase the Number of IPPs and Small and Medium Enterprises' in Their Legislative Objectives 131
4.16 Countries Citing 'Encourage Research' in Their Legislative Objectives 133
4.17 Countries Citing 'Increase Information About Renewable Energy/Public Education' in Their Legislative Objectives 134
4.18 Countries Citing 'Promote the Development of the Internal Energy Market and Regional Integration' in Their Legislative Objectives 136
4.19 Countries Citing 'Meet International Treaty Obligations and International Agreements' in Their Legislative Objectives 138
4.20 Countries Citing 'Environmental Protection' in Their Legislative Objectives 140
4.21 Countries Citing 'Reduce Greenhouse Gas Emissions and Address Climate Change' in Their Legislative Objectives 142
4.22 Countries Citing 'Support the Development of New Industry and Infrastructure' in Their Legislative Objectives 146
4.23 Countries Citing 'Encourage Technological Innovations' in Their Legislative Objectives 147
4.24 Countries Citing 'Create Jobs or Improve Skills and Domestic Capabilities' in Their Legislative Objectives 148
4.25 Countries Citing 'Local Manufacturing' in Their Legislative Objectives 151
4.26 Countries Citing 'Public Health or Improving Living Standards or Social Development' in Their Legislative Objectives 153
4.27 Countries Citing 'Access to Electricity' in Their Legislative Objectives 155
4.28 Countries Citing 'Affordable Energy' in Their Legislative Objectives 156
4.29 Countries Citing 'Promote Rural Development' in Their Legislative Objectives 158
5.1 Fundamental Types of Promotion Strategies 174

Acknowledgements

This book has benefited from the encouragement and support of a number of people. It is based on my doctoral thesis, which I completed at The University of Sydney Law School under the exceptional supervision and guidance of Professor Tim Stephens and Professor Terry Carney. Their understanding and insight as the research project morphed from studying three countries' laws to the laws of every applicable country in the world remains very much appreciated.

I would also like to thank my external examiners, Professor Adrian Bradbrook, Professor Angus Johnston and Professor Wang Mingyuan, for their careful examination and insightful feedback. This book has benefited from you so generously sharing your knowledge and expertise.

To my friends and family (which expanded over the course of writing this book), thank you for your caring support and understanding. I would particularly like to acknowledge the years of support provided by my parents, Max and Gail, and the very diligent proofreading provided by my mother. I am grateful for your help and it undoubtedly improved the book.

Thanks are also due to the editorial team at Cambridge University Press, Matt Gallaway, Jackie Grant, Catherine Smith and Dorothy Moyle who have been an absolute pleasure to work with. I would also like to acknowledge the excellent volunteer translation assistance provided by Ashley Richards, Stephanie Watson, Tallulah Bur, Mitchell Cleaver, Ellen Marie O'Brien, Levi Romanov and Laura Peck over the course of the project.

Finally, I thank Hayden and Madeleine, without whose steadfast love and support this book would not have been possible.

Acronyms and Abbreviations

AC	alternating current
$AU	Australian dollar
CDM	Clean Development Mechanism
CHP	combined heat and power
CO_2	carbon dioxide
CO_2e	carbon dioxide equivalent
COGEN	cogeneration
dB	decibel
DC	direct current
EC	European Commission
EIA	Energy Information Administration (US)
ES	energy storage
ETS	emissions trading scheme
EU	European Union
FIP	feed-in premium
FIT	feed-in tariff
GATT	General Agreement on Tariffs and Trade
G20	Group of Twenty
GDP	Gross Domestic Product
GHG	greenhouse gas
GW	gigawatt (1GW = 1000MW)
GWh	gigawatt-hour
GO	guarantee of origin
H_2O	water
IEA	International Energy Agency
IISD	International Institute for Sustainable Development
IPCC	Intergovernmental Panel on Climate Change
IPP	independent power producer

IPR	intellectual property right
IRENA	International Renewable Energy Agency
ITC	investment tax credit
kW	kilowatt (1kW = 1000watts)
kWh	kilowatt-hour
LNG	liquefied natural gas
LRET	large-scale renewable energy target
MW	megawatt (1MW = 1000kW)
MWh	megawatt-hour
NDRC	National Development and Reform Commission (China)
NFFO	Non-Fossil Fuel Obligation
NGO	non-governmental organisation
NREL	National Renewable Energy Laboratory (US)
NSW	New South Wales, Australia
OECD	Organisation for Economic Co-operation and Development
OPEC	Organization of Petroleum Exporting Countries
OTEC	ocean thermal energy conversion (also called 'maremotermica')
PM	particulate matter
PPA	power purchase agreement
PTC	production tax credit
PV	photovoltaic
R&D	research and development
REEEP	Renewable Energy and Energy Efficiency Partnership
RES	renewable energy sources
RES-E	electricity derived from renewable energy sources
RET	Renewable Energy Target
RO	Renewables Obligation
RPS	Renewable Portfolio Standards
SCM	Agreement on Subsidies and Countervailing Measures
SEDA	Sustainable Energy Development Authority of NSW (now defunct)
SOE	state-owned enterprises
SRET	small-scale renewable energy target
TFEU	Treaty on the Functioning of the European Union
TGC	tradeable green certificates
TPES	total primary energy supply
TW	terawatt (1TW = 1000GW)
UAE	United Arab Emirates
UK	United Kingdom
UN	United Nations

UNDP	United Nations Development Programme
UNEP	United Nations Environment Programme
UNFCCC	United Nations Framework Convention on Climate Change
UNIDO	United Nations Industrial Development Organization
US	United States of America
$US	United States dollar
VAT	value added tax
W	watt
WHO	World Health Organization
WTO	World Trade Organization

National Renewable Energy Legislation

Afghanistan

اساسنامه تصدی طرح تولید و ترویج وسایل استفاده از انرژی جدید و قابل تجدید 15/03/1369, [No. 101 Statute of the scheme means of production and promotion of new and renewable energy, 18 April 1989] [Linguistico Translations translation from Dari]

Albania

Për Burimet E Energjisë Së Rinovueshme No. 138/2013 [Law on Renewable Energy Sources 2013] [Linguistico Translations translation from Albanian]

Algeria

Loi n° 04–09 du 27 Joumada Ethania 1425 correspondant au 14 août 2004 relative à la promotion des énergies renouvelables dans le cadre du développement durable [Law No. 04–09 of 27 Jumada Ethania 1425 corresponding to 14 August 2004 on the promotion of renewable energies in the framework of sustainable development] [Stephanie Watson translation from French]

Andorra

Llei 93/2010, del 16 de desembre, de mesures de promoció de l'activitat econòmica i social, i de racionalització i d'optimització dels recursos de l'administració [Law 93/2010 of 16 December on the Promotion of Economic Activity and Social and Rationalisation and Optimisation of Resources Administration] [Linguistico Translations translation from Catalan]

Antigua and Barbuda

Renewable Energy Act 2015

Argentina

Energia Electrica Ley 26.190. Regimen de Fomento Nacional para el uso de fuentes renovables de energía destinada a la producción de energía eléctrica [Electricity Law 26.190: National Development Regimen for using renewable energy for electricity production 2006] [Linguistico Translations translation from Spanish]

Armenia

The Law of the Republic of Armenia on Energy Saving and Renewable Energy, No. ZR-67, 9 November 2004 [National Assembly of the Republic of Armenia translation from Armenian]

Australia

Renewable Energy (Electricity) Act 2000 (Cth)

Austria

75. Bundesgesetz über die Förderung der Elektrizitätserzeugung aus erneuerbaren Energieträgern (Ökostromgesetz) 2012 [75 Federal Act on the promotion of electricity generation from renewable energy sources (Green Electricity Act) 2012] [Mitchell Cleaver translation from German]

The Bahamas

Electricity Act 2015

Bangladesh

Sustainable and Renewable Energy Development Authority Act 2012 [Linguistico Translations translation from Bengali]

Barbados

Electric Light and Power Act 2013–21, 2014

Belarus

ЗАКОН РЕСПУБЛИКИ БЕЛАРУСЬ, 27 декабря 2010 г. № 204-З, О возобновляемых источниках энергии [Law of the Republic of Belarus, 27 December 2010 No. 204–3 on the use of Renewable Energy] [Levi Romanov translation from Russian]

Belgium

Arrêté royal relatif à l'établissement de mécanismes visant la promotion de l'électricité produite à partir des sources d'énergie renouvelables [Royal Decree on the establishment of mechanisms for the promotion of electricity produced from renewable energy sources, 16 July 2002] [Tallulah Bur translation from French]

Brazil

Law No. 10,438 of 2002 [Linguistico Translations translation from Portuguese]

Bulgaria

Закон за енергията от възобновяеми източници State Gazette No. 35/3.05.2011 [Energy from Renewable Sources Act 2011] [Bulgarian Government translation from Bulgarian]

Cabo Verde

Decreto-Lei No. 1/2011, promocao, incentivo, acesso, licenciamento, exploração, produção independente Energia Electrica [Decree Law No. 1/2011 on the promotion, access, licensing and development of electrical energy from renewable sources] [Linguistico Translations translation from Portuguese]

Cameroon

Law No. 2011/022 Governing the Electricity Sector in Cameroon

Chile

Ley núm. 20.257 introduce modificaciones a la ley general de servicios eléctricos respecto de la generación de energía eléctrica con fuentes de energías renovables no convencionales [Law No. 20,257 Non-Conventional Renewable Energy Law 2008] [Linguistico Translations translation from Spanish]

China

中华人民共和国可再生能源法 [Renewable Energy Law of the People's Republic of China 2005] [Ministry of Commerce of the People's Republic of China translation from Mandarin]

Colombia

Ley 697 de 2001 (Octubre 3) Diario Oficial No. 44.573, de 05 de octubre de 2001 mediante la cual se fomenta el uso racional y eficiente de la energía, se promueve la utilización de energías alternativas y se dictan otras disposiciones [Law 697 on the promotion of the efficient and rational use of energy and alternative energies, Official Gazette No. 44,573, 5 October 2001] [Linguistico Translations translation from Spanish]

Congo, Democratic Republic of the

Loi n° 14/011 du 17 juin 2014 relative au secteur de lélectricité [Law No. 14/011 of 17 June 2014 relating to the Electricity Sector] [Linguistico Translations translation from French]

Cote d'Ivoire
The Code of Electricity [The Electricity Code, Law No. 2014–132]

Croatia
Zakon o obnovljivim izvorima energije i visokoučinkovitoj kogeneraciji [Law on Renewable Energy Sources and Highly Effective Cogeneration, No. 100/2015] [Croatian Government translation from Croatian]

Cuba
Decreto Ley 345: Una revolución dentro de la revolución energética [Law Decree 345 on the development of renewable resources and efficient use of energy 2018] [Linguistico Translations translation from Spanish]

Cyprus
Νομοσ που προνοει για την προωθηση και ενθαρρυνση τησ χρησησ
Των ανανεωσιμων πηγων ενεργειασ [The Promotion and Encouragement of the Use of Renewable Energy Act of 2013] [Cyprus Energy Regulatory Agency translation from Greek]

Czech Republic
Zákon č. 165/2012 Sb. podporovaných zdrojích energie [Act No. 165/2012 on promoted energy sources] [Linguistico Translations translation from Czech]

Denmark
Lov om fremme af vedvarende energi [Promotion of Renewable Energy Act, No. 1392 of 2008] [Global Denmark Translations on behalf of the Danish Government translation from Danish]

Dominican Republic
Ley No. 5707 sobre Incentivo al Desarrollo de Fuentes Renovables de Energía y de sus Regímenes Especiales [Renewable Energies Incentive Law 57-07, 2007] [Linguistico Translations translation from Spanish]

Ecuador
Ley Orgánica del Servicio Público de Energía Eléctrica [Organic Law on the Public Service of Electricity, Electricity Law 2015] [Linguistico Translations translation from Spanish]

Egypt, Arab Republic of
قرار رئيس الجمهورية بالقانون رقم 203 لسنة 2014 بشأن تحفيز إنتاج الكهرباء من مصادر الطاقة المتجددة [Decree-Law of the President of the Arab Republic of Egypt No. 203 of 2014 Regarding the Stimulation of Producing Electricity from

Renewable Energy Sources] [The Middle East Library for Economic Services translation from Arabic]

El Salvador

Ley De Incentivos Fiscales Para El Fomento De Las Energías Renovables En La Generación De Electricidad [Fiscal Incentives Law for the Promotion of Renewable Energy, No. 462, 2007] [Linguistico Translations translation from Spanish]

Estonia

Elektrituruseadus [Electricity Market Act 2003] [Estonian Government translation from Estonian]

Finland

Laki uusiutuvilla energialähteillä tuotetun sähkön tuotantotuesta [Act on Production Subsidy for Electricity Produced from Renewable Energy Sources, Act No. 1396/2010] [Finlex (Ministry of Justice, Government of Finland) translation from Finnish]

France

Code de l'énergie, version consolidée au 1 août 2018 [Energy Code, version consolidated to 1 August 2018] [Legifrance (Government of France) translation from French]

The Gambia

The Renewable Energy Act 2013

Germany

Gesetz für den Ausbau erneuerbarer Energien (Erneuerbare-Energien-Gesetz – EEG 2017) [Renewable Energy Sources Act 2017] [German Federal Ministry for Economic Affairs and Energy translation from German]

Ghana

Renewable Energy Act 2011

Greece

Νόμος 4414/2016 Νέο καθεστώς στήριξης των σταθμών παραγωγής ηλεκτρικής ενέργειας από Ανανεώσιμες Πηγές Ενέργειας και Συμπαραγωγή Ηλεκτρισμού και Θερμότητας Υψηλής Απόδοσης – Διατάξεις για το νομικό και λειτουργικό διαχωρισμό των κλάδων προμήθειας και διανομής στην αγορά του φυσικού αερίου και άλλες διατάξεις [Law No. 4414/2016 New Support Scheme of Renewable Energy and CHP Plants – Provisions concerning the Legal and

Administrative Unbundling of Natural Gas Supply and Distribution and Miscellaneous Provisions] [Greek Government translation from Greek]

Guatemala

Decreto número 52–2003 ley de incentivos para el desarrollo de proyectos de energia renovable 2003 [Renewable Energy Project Incentives Act 2003] [Linguistico Translations translation from Spanish]

Honduras

Ley de Promocion a la Generacion de Energia Electrica con Recursos Renovables 2007 [Law for the Promotion of Electricity Generation with Renewable Resources 2007] [Linguistico Translations translation from Spanish]

Hungary

2007. évi LXXXVI. törvény a villamos energiáról [Act No. LXXXVI of 2007 on Electric Energy] [Linguistico Translations translation from Hungarian]

Iceland

Raforkulaga [Electricity Act 2003] [Orkustofnun (National Energy Authority) translation from Icelandic]

India

The Electricity Act 2003

Indonesia

Undang-Undang Republik Indonesia Nomor 30 Tahun 2007 Tentang Energi [Law of the Republic of Indonesia Number 30 of 2007 About Energy] [Ellen Marie O'Brien translation from Indonesian]

Ireland

Electricity Regulation Act 1999

Italy

Decreto Legislativo 3 marzo 2011, n. 28 Attuazione della direttiva 2009/28/CE sulla promozione dell'uso dell'energia da fonti rinnovabili, recante modifica e successive abrogazione delle direttive 2001/77/CE e 2003/30/CE [Legislative Decree No. 28, 2011 Implementation of the Directive 2009/28/EC on the promotion of the use of energy from renewable sources] [Linguistico Translations translation from Italian]

Jamaica

The Electricity Act 2015

Japan

電気事業者による再生可能エネルギー電気の調達に関する特別措置法 [Act on Special Measures Concerning Procurement of Electricity from Renewable Energy Sources by Electric Utilities, No. 108 of 2011] [Ministry of Justice translation from Japanese]

Jordan

قانون الطاقة المتجددة وكفاءة الطاقة رقم 13 لسنة 2012 [The Renewable Energy and Energy Efficiency Law, Law No. 13 of 2012] [Kingdom of Jordan translation from Arabic]

Kazakhstan

О поддержке использования возобновляемых источников энергии Закон Республики Казахстан от 4 июля 2009 года № 165-IV [Law of the Republic of Kazakhstan No. 165-IV About the Support of the Use of Renewable Energy of 4 July 2009] [Government of the Republic of Kazakhstan translation from Kazakh]

Kenya

The Energy Act 2006

Korea, South

신에너지 및 재생에너지 개발・이용・보급 촉진법 [Act on the Promotion of the Development, Use and Diffusion of New and Renewable Energy, No. 14670, 2017] [Korean Legislative Research Institute translation from Korean]

Kosovo

Law on Energy 2016

Kyrgyzstan

Закон Кыргызской Республики "О Возобновляемых Источниках Энергии [Law of Kyrgyz Republic 'On Renewable Energy', No. 283 of 2008] [Levi Romanov translation from Russian]

Latvia

Elektroenerģijas tirgus likums 2005 [Electricity Market Law 2005] (Latvia) 82 115/0825/201105 [Latvian Government translation from Latvian]

Liechtenstein

Gesetz vom 24. April 2008 über die Förderung der Energieeffizienz und der erneuerbaren Energien (Energieeffizienzgesetz; EEG) [Energy Efficiency Act (24 April 2008) Law on the requirement of energy efficiency and renewable energy (EEG)] [Laura Peck translation from German]

Lithuania

Atsinaujinančių išteklių energetikos įstatymas [Law on Energy from Renewable Sources No. XI-1375, 2011] [Linguistico Translations translation from Lithuanian]

Luxembourg

Loi du 18 février 2010 relative à un régime d'aides à la protection de l'environnement et à l'utilisation rationnelle des ressources naturelles [Act of 18 February 2010 establishing a support system for environmental protection and rational use of natural resources] [Tallulah Bur translation from French]

Macedonia

Закон За Енергетика [Energy Law 2018] [Government of the Republic of Macedonia translation from Macedonian]

Madagascar

Loi n° 2017–020 portant Code de l'Electricité à Madagascar [Law No. 2017–020 Madagascan Electricity Code] [Linguistico Translations translation from French]

Malawi

Energy Regulation Act 20 of 2004

Malaysia

Renewable Energy Act 2011

Malta

Skema Ta' Tariffi Feed-In (Elettriku Ġenerat Minn Stallazzjonijiet Fotovoltajċi Tax-Xemx) (Legislazzjoni Sussidjarja 545.27 Regolamenti Dwar Skema Ta' Tariffi Feed-In) [Feed-In Tariffs Scheme (Electricity Generated from Solar Photovoltaic Installations) 2013] [Linguistico Translations translation from Maltese]

Mauritius

The Mauritius Renewable Energy Agency Act 2015

Mexico

Ley De Transición Energética [Energy Transition Law 2015] [Linguistico Translations translation from Spanish]

Moldova

Закон Республики Молдова От 26 Февраля 2016 Года №10 О Продвижении Использования Энергии Из Возобновляемых

Источников [Law of the Republic of Moldova of February 26, 2016 No. 10 About promotion of energy use from renewable sources] [Linguistico Translations translation from Russian]

Mongolia

Монгол Улсын Хууль Сэргээгдэх Эрчим Хүчний Тухай [Law of Mongolia on Renewable Energy 2007] [Ministry of Mineral Resources and Energy, Mongolia translation from Mongolian]

Montenegro

Zakon o energetici [Energy Law, Law as published in the Official Gazette of Montenegro, No. 5/2016 and 51/2017] [Montenegro Investment Promotion Agency translation from Montenegrin]

Morocco

Loi n° 13–09 relative aux énergies renouvelables [Renewable Energy Law No. 13.09, 2010] [Tallulah Bur translation from French]

Netherlands

Electricity Act 1998 [Government of the Netherlands translation from Dutch]

Nicaragua

Ley Para La Promoción De Generación Eléctrica Con Fuentes Renovables [Law for the Promotion of Renewable Energy Generation, No. 532, 2005] [Nicaraguan Government translation from Spanish]

Norway

Lov om elsertifikater 2011 [Electricity Certificates Act 2011] [Linguistico Translations translation from Norwegian]

Pakistan

Regulation of Generation, Transmission and Distribution of Electric Power Act, Act XL of 1997

Palau

Palau Energy Act of 2015, RPPL 9–54, amending Title 37 of the Palau National Code

Palestine

قرار بقانون رقم 14 لسنة 2015 بشأن الطاقة المتجددة وكفاءة الطاقة [Decree Law No. 14/2015 on renewable energy and energy efficiency] [Birzeit University Institute of Law translation from Arabic]

Panama

Asamblea Legislativa Ley No. 6 Por La Cual Se Dicta El Marco Regulatorio E Institucional Para La Prestacion Del Servicio Publico De Electricidad [Legislature Law No. 6 The Regulatory and Institutional Framework for the Provision of Public Electricity Services 1997] [Linguistico Translations translation from Spanish]

Paraguay

Poder Legislativo Ley No. 3009 De La Producción Y Transporte Independiente De Energía Eléctrica (PTIEE) [Law No. 3009 On the Production and Transport of Independent Electricity (PTIEE) 2006] [Linguistico Translations translation from Spanish]

Peru

Decreto Legislativo De Promoción De La Inversión Para La Generación De Electricidad Con El Isuo De Energías Renovables [Legislative Decree of Investment Promotion for Electricity Generation with the Use of Renewable Energy, No. 1002/2008] [Linguistico Translations translation from Spanish]

Philippines

Renewable Energy Act of 2008

Poland

USTAWA z dnia 20 lutego 2015 r. o odnawialnych źródłach energii [Law on renewable energy sources, 20 February 2015] [Poland Office of Sejm translation from Polish]

Portugal

Ministry of Industry and Energy Decreto-Lei n.° 141/2010 [Decree-Law No. 141/2010] [Linguistico Translations translation from Portuguese]

Romania

Legea 220/2008 pentru stabilirea sistemului de promovare a producerii energiei din surse regenerabile de energie, republicata 2010 [Law 220/2008 on establishing the promotion system of energy production from renewable energy sources] [Linguistico Translations translation from Romanian]

Russia

ПОСТАНОВЛЕНИЕ От 28 Мая 2013 Г. № 449 О Механизме Стимулирования Использования Возобновляемых Источников Энергии На Оптовом Рынке Электрической Энергии И Мощности

[Decree of 28 May 2013 No. 449 on the mechanism for promoting the use of renewable energy sources in the wholesale market of electric energy and power] [CIS Legislation translation from Russian]

San Marino

Legge 7 maggio 2008 n.72 – Promozione ed incentivazione dell'efficienza energetica degli edifici e dell'impiego di energie rinnovabili in ambito civile e industriale [Law No. 72 on the Promotion and Incentives for Energy Efficiency in Buildings and the Use of Renewable Energy in the Civil and Industrial Fields 2008] [Linguistico Translations translation from Italian]

Senegal

Loi n° 2010–21 du 20 Décembre 2010 relative à l'orientation sur le droit de l'énergie renouvelable [Act No. 2010–21 of 20 December 2010 concerning guidance on renewable energy law] [Ashley Richards translation from French]

Serbia

Закон о енергетици No. 145/2014 [Energy Law 2014] [Energy Agency, Republic of Serbia translation from Serbian]

Seychelles

Energy Act 2012

Slovakia

Zákon 309/2009 Z.z. o podpore obnovitelných zdrojov energie [Act 309 of 19 June 2009 on the promotion of renewable energy sources and high-efficiency cogeneration and on amendments to certain acts] [Slovakian Government translation from Slovakian]

Slovenia

Energetski zakon No. 17/14, 2014 [Energy Act 2014] [Linguistico Translations translation from Slovenian]

South Africa

National Energy Act 2008

Spain

Real Decreto 413/2014 de 6 de junio, por el que se regula la actividad de producción de energía eléctrica a partir de fuentes de energía renovables, cogeneración y residuos [Royal Decree 413/2014 regulating the activity of electricity production from renewable energy sources, cogeneration and waste] [Spanish Government translation from Spanish]

Sri Lanka
Sustainable Energy Authority Act 2007

St. Lucia
Chapter 9.02 Electricity Supply Act 1994

St. Vincent and the Grenadines
Geothermal Resources Development Act 2015

Suriname
Elektriciteitswet 2016 [Electricity Act 2016] [Linguistico Translations translation from Dutch]

Sweden
Lag om elcertifikat No. 2011:1200 [Electricity Certificates Act 2011] [Swedish Government translation from Swedish]

Switzerland
SR 730.0 Energiegesetz EnG 2016 [Energy Act 2016, SR 730.0] [Swiss Federal Office of Energy translation from German]

Syria
قانون الكهرباء [Electricity Law, No. 32 of 2010] [Linguistico Translations translation from Arabic]

Taiwan
再生能源發展條例 [Renewable Energy Development Act 2009] [Laws & Regulations Database of The Republic of China translation from Mandarin]

Tajikistan
Закон Республики Таджикистан Об Использовании Возобновляемых Источников Энергии [The Law of the Republic of Tajikistan on the use of Renewable Energy Sources 2010] [UNDP in Tajikistan translation from Russian]

Thailand
พระราชบัญญัติ การประกอบกิจการพลังงาน พ.ศ. ๒๕๕๐ [Energy Industry Act 2007, B E 2550] [Thai Law Forum Translation from Thai]

Togo
Loi No. 2018–010 relative a la promotion de la production d'électricité à base des sources d'énergies renouvelables [Clean Energy Act 2018] [Linguistico Translations translation from French]

Tonga

Renewable Energy Act 2008

Tunisia

Loi n° 2015–12 du 11 mai 2015, relative à la production d'électricité à partir des énergies renouvelables [Law No. 2015–12 dated 11 May 2015 relating to the electricity generation from renewable energies] [Ashley Richards translation from French]

Turkey

Yenilenebilir Enerji Kaynaklarinin Elektrik Enerjisi Üretimi Amaçli Kullanimina İlişkin Kanun Kanun Numarası [Law Regarding the Use of Renewable Energy Resources for Electricity Production, No. 5346, 2005] [Linguistico Translations translation from Turkish]

Ukraine

Закон України Про альтернативні джерела енергії [Law on Alternative Energy Sources, No. 555-IV, 2003] [Linguistico Translations translation from Ukrainian]

United Kingdom

Electricity Act 1989

Energy Act 2008

The Promotion of the Use of Energy from Renewable Sources Regulations 2011

United States of America

Public Utilities Regulatory Policies Act 1978

Energy Policy Act of 2005

Uruguay

Decreto N° 354/009 se declaran promovidas las actividades tendientes a la generación de energía eléctrica 2009 [Declaration No. 354/009 promoting activities aimed at generating electricity 2009] [Linguistico Translations translation from Spanish]

Uzbekistan

Energiyadan oqilona foydalanish to'g'risida [Law on the Rational Use of Energy 1997, No. 412-I,] [Uzbekistan Energy Centre translation from Uzbek]

O'zbekiston Respublikasi Vazirlar Mahkamasining 2015 yil 13 avgustdagi № 238-son Qarori bilan energiya samaradorpligi va qayta tiklanadigan energiya manbalarini rivojlantirish masalalari bo'yicha Respublika komissiya

[Resolution of the Cabinet of Ministers of the Republic of Uzbekistan of 13 August 2015 No. 238 about approval of the Regulations on the Republican Commission on questions of energy efficiency and development of renewable energy resources] [CIS Legislation translation from Uzbek]

Venezuela, Republic of

Ley de uso racional y eficiente de la energía [Law on the Rational and Efficient Use of Energy 2011] [Linguistico Translations translation from Spanish]

Yemen

قانون الكهرباء رقم 1 لسنة 2009 [Electricity Law 2009] [Linguistico Translations translation from Arabic]

Other Legislation and Treaties

Acts Interpretation Act 1901 (Australia)

Agreement on Subsidies and Countervailing Measures, WTO Doc 1869 UNTS 14 (15 April 1994)

Carbon Tax Act 2008 (British Columbia)

Code Général des Impôsts, version consolidée au 1 septembre 2018 [General Tax Code, version consolidated to 1 September 2018] [Legifrance (Government of France) translation from French]

Council of Ministers Decision No. 2013/5625 published in the Official Gazette No. 28842, dated 5 December 2013 (Turkey)

VYHLÁŠKA N°18/2017 Úradu pre reguláciu sieťových odvetví z 8. februára 2017, ktorou sa ustanovuje cenová regulácia v elektroenergetike a niektoré podmienky vykonávania regulovaných činností v elektroenergetike [DECREE N° 18/2017 of the Regulatory Office for Network Industries dated 8 February 2017 establishing price regulation in the electricity sector and certain conditions for the implementation of regulated activities in the electricity sector] (Slovakia)

Directive 2001/77/EC of the European Parliament and of the Council of 27 September 2001 on the promotion of electricity produced from renewable energy sources in the internal electricity market [2001] OJ L 283

Directive 2009/28/EC of the European Parliament and of the Council of 23 April 2009 on the promotion of the use of energy from renewable sources and amending and subsequently repealing Directives 2001/77/EC and 2003/30/EC [1993] OJ L 140/16

Electricity Supply Act 1995 (NSW)

Electricity Supply Amendment (Solar Bonus Scheme) Act 2009 (NSW)

Environmental Pollution Fee 1999 (Ukraine)

International Renewable Energy Agency, Statute of the International Renewable Energy Agency (Adopted at the Conference on the Establishment of the International Renewable Energy Agency, Bonn, 26 January 2009, entered into force 8 July 2010)

Νόμος 4399/2016 – Νόμος 4399/2016 Θεσμικό πλαίσιο για τη σύσταση καθεστώτων Ενισχύσεων Ιδιωτικών Επενδύσεων για την περιφερειακή και οικονομική ανάπτυξη της χώρας – Σύσταση Αναπτυξιακού Συμβουλίου και άλλες διατάξεις [Law 4399/2016 Institutional Framework for the Establishment of Private Investment Aid Schemes for the Regional and Economic Development of the Country – Establishment of a Development Council and other provisions] (Greece) [Start-up Greece (Government of Greece) translation from Greek]

风力发电设备产业化专项资金管理暂行办法 [Management Regulations on Special Fund for the Industrialization of Wind Power Manufacturing Sector in China] (Ministry of Finance Document No. 476, People's Republic of China, 11 August 2008)

Marrakesh Agreement Establishing the World Trade Organization, opened for signature 15 April 1994, [1994] 1867 UNTS 3 (entered into force 1 January 1995)

Renewable Energy Development Act 2009 (Utah)

Renewable Energy (Electricity) Amendment Act 2015 (Australia)

Renewable Energy (Electricity) Regulations 2001 (Australia)

Resolution of the Council of Ministers of the Republic of Belarus of 07.08.2015 N°45 'On the tariffs for electricity produced from renewable energy sources on the territory of the Republic of Belarus by individual entrepreneurs and legal entities who are not members of the State Electricity Production Association "Belenergo", and released to supply companies of this association'

Treaty on European Union, opened for signature 7 February 1992, [2009] OJ C 115/13 (entered into force 1 November 1993)

Treaty on the Functioning of the European Union, opened for signature 7 February 1992, [2009] OJ C 115/13 (entered into force 1 November 1993)

United Nations Convention on the Law of the Sea, opened for signature 10 December 1982, 1833 UNTS 3 (entered into force 16 November 1994)

Valtioneuvoston asetus uusiutuvan energian ja uuden energiateknologian investointituen myöntämisen yleisistä ehdoista 25.02.2016/145 [Government Decree No. 145/2016 on Granting Investment Aid for Renewable Energy and New Energy Technologies] (Finland)

Units of Measurement

Throughout this book, reference is made to both units of power and units of energy. A brief explanation of the difference between these two common units of measurement is provided below.

Units of Power

The watt (W) is a unit of power, which represents the *rate* at which energy is produced or consumed at a specific point in time. For example, light bulbs are designated into categories by watts. A 100W light bulb will require a greater flow of energy to power the globe than a 30W light bulb.

Units of Energy

The watt-hour (Wh) is a unit of energy, which represents the *amount* of energy used or generated to carry out 'work' within a specified time period, in this case an hour. For example, a kilowatt-hour (kWh) is the amount of energy used in one hour at a rate of 1,000 watts.

Standard Prefixes for Units of Power and Units of Energy

The following prefixes are used in relation to both watts and watt-hours:
kilo (k) = 1,000 $=10^3$
Mega (M) = 1,000,000 = 10^6
Giga (G) = 1,000,000,000 = 10^9
Tera (T) = 1,000,000,000,000 = 10^{12}
Peta (P) = 1,000,000,000,000,000 = 10^{15}

1

Introduction

Over the past fifteen years, the renewable energy sector has experienced an unprecedented boom. In 2017, renewable energy accounted for an estimated 70 per cent of net additions to global power generation,[1] with $US298 billion invested in new renewable generation capacity (including hydropower).[2] This continues the trend witnessed over the past five years of renewables attracting more than double the annual investment into new fossil fuel generation capacity.[3] Nor does the boom show any sign of abating. As part of the adoption of the Paris Agreement, 145 Parties included domestic action to support renewable energy to help mitigate and adapt to climate change as part of their Nationally Determined Contributions.[4] Further, 109 Parties to the Paris Agreement provided quantifiable targets for renewables. These developments are having significant flow-on effects to the global generation mix, with renewable energy now comprising 26.5 per cent of global power generation capacity from all sources and delivering 25 per cent of global electricity supply.[5] This growth reflects the role that governments believe renewable energy will play in ensuring national energy security, combating climate change and sustainably meeting rising energy demands.

Despite this rapid growth, substantially more investment is needed if projections from the International Energy Agency (IEA) of an almost tripling

1 REN21 Secretariat, 'Renewables 2018 Global Status Report' (Report, Renewable Energy Policy Network for the 21st Century, 2018) 18.

2 IEA, *World Energy Investment 2018* (OECD/IEA, 2018) 23.

3 Ibid.

4 IRENA, *Untapped Potential for Climate Action: Renewable Energy in Nationally Determined Contributions* (IRENA, 2017) Executive Summary.

5 REN21 Secretariat, above n 1, 32, 40.

of renewable generation by 2040 are to be met.[6] Indeed, the IEA has predicted that 'cumulative investment of $US 7.8 trillion is needed for renewable energy supply in the period to 2040, around 95% of which should be spent on power generation technologies'.[7] The International Renewable Energy Agency (IRENA) puts the level of cumulative investment needed to enable the energy transition far higher, estimating that $US120 trillion will need to be spent between 2015 and 2050, with the vast majority of the spending targeted at renewable energy and energy efficiency.[8] Regardless of the figure, both the IEA and IRENA agree that significantly more investment is needed within the sector. To date, global investment in the renewable energy sector has been highly variable year on year, although there has been a trend towards upwards growth. However, due to a confluence of factors, the precise investment in any year can be unpredictable, with, for example, investment in 2016 some 23 per cent lower than the global peak in investment in 2015.[9]

The downturn in renewable energy investment in 2016 is not a new phenomenon, with similar declines experienced in 2012, 2013 and 2014 from the previous global peak in investment in 2011.[10] Historically, the variable nature of global investment in the renewable energy sector can be attributed to several factors such as:

- Policy and regulatory support: The renewables sector needs policy and regulatory certainty in order to be able to function efficiently, plan investments to avoid boom and bust cycles, and to enable the energy transition. The expiry of the green infrastructure stimulus packages following the end of the global financial crisis[11] and the subsequent increased uncertainty about national renewable energy laws and policy following the early closure or retroactive declines in support schemes in some countries[12] has had a marked impact on investment in the sector. This has been shown by the huge growth in investor–state disputes and

6 See the Sustainable Development Scenario in IEA, *World Energy Outlook* (OECD/IEA, 2017) 257, 299.

7 IEA, Renewable Energy Outlook in IEA, *World Energy Outlook* (OECD/IEA, 2014) 239.

8 IRENA, *Global Energy Transformation: A Roadmap to 2050* (IRENA, 2018) 11.

9 REN21 Secretariat, above n 1, 140.

10 Angus McCrone (ed.) et al., *Global Trends in Renewable Energy Investment 2017* (Frankfurt School-UNEP Centre and Bloomberg New Energy Finance, 2017) 14.

11 Ibid.

12 Thomas Gerke, *Italy Imposes Retroactive Changes to Feed-in Tariff for Solar PV*, Renew Economy, 15 August 2014 <http://reneweconomy.com.au/2014/italy-imposes-retroactive-changes-feed-tariff-pv-38857>; Nilima Choudhury, *Spain Announces Retroactive FiT Cuts*, PV Tech, 19 February 2013 <www.pv-tech.org/news/spain_announces_retroactive_fit_cuts>.

associated compensation claims within the sector, particularly in Spain, the Czech Republic, Bulgaria and Italy.[13]

- Technology and equipment costs: The average capital cost of renewables projects has declined markedly as onshore wind and photovoltaic solar became increasingly cost competitive with fossil fuel generation projects. Since 2009, photovoltaic solar module prices have declined by approximately 80 per cent and wind turbine equipment costs have fallen by 30 to 40 per cent.[14] In turn, this led to a decline in the levelised cost of electricity (LCOE) from solar PV between 2010 and 2016 of more than 68 per cent, with an 18 per cent decline in the LCOE of onshore wind over the same period.[15] This is obviously a positive development, though the benefit of falling costs will only be realised if there is sufficient regulatory certainty to encourage investment.
- Fossil fuel costs: A further factor that has impacted investment within the sector in recent years is the comparatively low cost of fossil fuels, following large discoveries of shale gas and oil in the United States.[16] In 2016, US gasoline prices dropped to levels not seen since 2004.[17] This meant that the cost imperative to substitute fossil fuelled generation with renewable energy, which had been present when oil prices reached their historic peak in 2014,[18] was no longer as pressing. While global oil and gas prices have now largely recovered, the linkages between the prices of different sources of electricity generation can lead to the attractiveness of investing in the renewables sector being impacted by other factors. This means that international fossil fuel commodity prices or conflicts in the Middle East and Russia continue to impact on investment in the renewable energy sector.

The renewable energy sector is also affected by the presence of at least three market failures within the energy sector: the presence of unpriced negative and positive externalities in the energy sector; spillovers and learning effects; and information asymmetries. These market failures, which afflict countries

13 See e.g. Energy Charter Treaty, *List of All Investment Dispute Settlement Cases* (2018) <https://energycharter.org/what-we-do/dispute-settlement/all-investment-dispute-settlement-cases/>.

14 IEA and IRENA, *Perspectives for the Energy Transition: Investment Needs for a Low-Carbon Energy System* (OECD/IEA and IRENA, 2017) 27.

15 IRENA, *Renewable Energy: A Key Climate Solution* (IRENA, 2017) 5.

16 European Renewable Energy Council, 'Shale Gas and Its Impact on Renewable Energy Sources' in EREC (ed.), *EREC Factsheet* (2013).

17 Trading Economics, *United States Gasoline Prices 1992–2018* (2018) <https://tradingeconomics.com/united-states/gasoline-prices>.

18 Ibid.

all over the world, are further combined with a range of market barriers that vary by country. Arguably, the most significant market barrier is the ongoing subsidies provided to fossil fuel and nuclear generation, with direct subsidies provided to fossil fuels (i.e. excluding nuclear generation) reaching $US260 billion in 2016 alone[19] (or 0.34 per cent of global gross domestic product (GDP)).[20] This may be compared to the direct subsidies provided to renewable energy used for power generation over the same period of $US140 billion.[21] These factors mean that without government intervention in the sector, many sources of renewable energy cannot effectively compete with fossil fuel generation and would not have a sufficient market, price or profitability potential to warrant improving existing technologies, reducing their costs and the development of new technologies.[22]

As a result, a majority of the countries in the world engage in some form of government intervention in the renewable energy sector, with the numbers consistently growing year on year. By 2018, 146 countries had national renewable power targets,[23] 138 countries had support policies directed at renewable energy[24] and 113 countries had national renewable energy laws in force. In addition, many countries have developed other flexible policy mechanisms to encourage the development and commercialisation of new renewable generation. These mechanisms include research and development (R&D) support, reforms to planning laws and improving key market infrastructure such as the reinforcement of the electricity transmission and distribution networks so that they can cope with increased loads and intermittency of supply.

1.1 THE PROBLEM AND SIGNIFICANCE OF THE RESEARCH

Given the considerable growth of the renewable energy sector and the frequency with which governments intervene to support its ongoing development,

19 IEA, *World Energy Outlook*, above n 6, 85.

20 Author's own calculations using the global gross domestic product data for 2016 from World Bank, *World Development Indicators database* (1 July 2018) <https://databank.worldbank.org/data/download/GDP.pdf>.

21 IEA, *World Energy Outlook*, above n 6, 82.

22 See e.g. Department of Energy & Climate Change, *UK Renewable Energy Roadmap Update 2013* (Government of the United Kingdom, 2013) 5; Australian Renewable Energy Agency, *The Business of Renewables: A Report into Renewable Energy Take-up by Large Corporates in Australia* (ARENA, 2017) 44–5; Ernst & Young, *Capitalizing on China's Renewable Energy Opportunities: Innovative Financing Models for China's Solar and Wind Markets* (Ernst & Young (China) Advisory, 2014) 4.

23 REN21 Secretariat, above n 1, Table R8.

24 Ibid.

it is surprising that there has not been a comprehensive scholarly analysis of the national renewable energy law for every country that has such a law. This means that fundamental issues such as how renewable energy is defined in law, what countries are trying to achieve through their national renewable energy laws and how they combine regulatory support mechanisms to achieve this, have often been neglected in research.

Much of the prior research into the renewable energy sector has been focused on the relative efficiency and efficacy of the different regulatory models available.[25] These studies have analysed whether feed-in tariffs, quota systems such as renewable portfolio standards (RPS), competitive tendering, tax incentives, subsidies or other regulatory support mechanisms are preferable.[26] Due to the focus of these studies on relative efficiency and effectiveness, much of this work has adopted a strong economic focus and/or policy orientation.[27] Further, the existing multi-country comparisons in this area have tended to focus on Europe,[28]

[25] See e.g. Michael B Gerrard (ed.), *The Law of Clean Energy: Efficiency and Renewables* (American Bar Association Section of Environment, Energy, and Resources, 2012); Jonathan A Lesser and Xuejuan Su, 'Design of an Economically Efficient Feed-in Tariff Structure for Renewable Energy Development' (2008) 36 *Energy Policy* 981; Richard L Ottinger and Adrian J Bradbrook (eds.), *UNEP Handbook for Drafting Laws on Energy Efficiency and Renewable Energy Resources* (UNEP/ Earth Print Limited, 2007); Maria Ellingson et al., *Compendium of Best Practices: Sharing Local and State Successes in Energy Efficiency and Renewable Energy from the United States* (REEEP/ACORE, 2010).

[26] Peng Sun and Pu-yan Nie, 'A Comparative Study of Feed-in Tariff and Renewable Portfolio Standard Policy in Renewable Energy Industry' (2015) 74 *Renewable Energy* 255; C G Dong, 'Feed-in Tariff vs. Renewable Portfolio Standard: An Empirical Test of Their Relative Effectiveness in Promoting Wind Capacity Development' (2012) 42 *Energy Policy* 476; Reinhard Haas et al., 'A Historical Review of Promotion Strategies for Electricity from Renewable Energy Sources in EU Countries' (2011) 15 *Renewable and Sustainable Energy Reviews* 1003, 1026; Lucy Butler and Karsten Neuhoff, 'Comparison of Feed-in Tariff, Quota and Auction Mechanisms to Support Wind Power Development' (2008) 33 *Renewable Energy* 1854, 1858; Toby Couture and Yves Gagnon, 'An Analysis of Feed-in tariff Remuneration Models: Implications for Renewable Energy Investment' (2010) 38 *Energy Policy* 955, 955.

[27] See e.g. Lesser and Su, 'Design of an Economically Efficient Feed-in Tariff Structure', above n 25; Severin Borenstein, 'The Private and Public Economies of Renewable Electricity Generation' 26 *The Journal of Economic Perspectives* 67.

[28] See e.g. Dörte Fouquet, 'Policy Instruments Renewable Energy – From a European Perspective' (2013) 49 *Renewable Energy* 15; António C Marques, José A Fuinhas and J R Pires Manso, 'Motivations Driving Renewable Energy in European Countries: A Panel Data Approach' (2010) 38 *Energy Policy* 6877; Pablo del Río et al., 'Key Policy Approaches for a Harmonisation of RES(-E) Support in Europe – Main Options and Design Elements' (Report, European IEE Project Beyond2020, March 2012); Lena Kitzing, Catherine Mitchell and Poul Erik Morthorst, 'Renewable Energy Policies in Europe: Converging or Diverging?' (2012) 51 *Energy Policy* 192; Gustav Resch et al., 'Coordination or Harmonisation? Feasible Pathways for a European RES Strategy Beyond 2020' (2013) 24

North America,[29] Organisation for Economic Co-operation and Development (OECD) countries[30] or north-east Asia.[31] This prompts the question of whether the outcomes of this research are actually generalisable across all countries with such laws, and developing nations in particular.

This book seeks to address this gap in the literature. It is the first research to analyse the primary piece of national renewable energy legislation from each of the 113 countries that had such a law on 1 August 2018, as well as the EU Directive[32]

Energy and Environment 147; David Jacobs, *Renewable Energy Policy Convergence in the EU: The Evolution of Feed-in Tariffs in Germany, Spain and France* (Ashgate Publishing, 2012); Sian Crampsie, 'Renewables convergence?' (2011) 34(14) *Utility Week* 9; Tatiana Romanova, 'Legal Approximation in Energy: A New Approach for the European Union and Russia' in Caroline Zuzemko, Andrei V Belyi, Andreas Goldthau and Michael F Keating (eds.), *Dynamics of Energy Governance in Europe and Russia* (Palgrave Macmillan, 2012) 23; Miquel Muñoz, Volker Oschmann and J David Tàbara, 'Harmonization of Renewable Electricity Feed-in Laws in the European Union' (2007) 35 *Energy Policy* 3104; Roger Hildingsson, Johannes Stripple and Andrew Jordan, 'Governing Renewable Energy in the EU: Confronting a Governance Dilemma' (2012) 11 *European Political Science* 18; Malgorzata Alicja Czeberkus, *Renewable Energy Sources: EU Policy and Law in Light of Integration* (LLM Thesis, University of Iceland, 2013); Per-Olof Busch and Helge Jörgens, 'Europeanization Through Diffusion? Renewable Energy Policies and Alternative Sources for European Convergence' in Francesc Morata and Israel Solorio Sandoval (eds.), *European Energy Policy* (Edward Elgar, 2012) 66.

29 Thomas P. Lyon and Haitao Yin, 'Why Do States Adopt Renewable Portfolio Standards? An Empirical Investigation' (2010) 31 *The Energy Journal* 131; Clean Energy States Alliance, *Developing an Effective State Clean Energy Program: Competitive Grants* (CESA, 2009); Maria Ellingson et al., above n 25; Steffen Jenner et al., 'What Drives States to Support Renewable Energy?' (2012) 33(2) *Energy Journal* 1; Warren Leon and Clean Energy States Alliance, 'Designing the Right RPS: A Guide to Selecting Goals and Program Options for a Renewable Portfolio Standard' (Guide, State-Federal RPS Collective and the National Association of Regulatory Utility Commissioners, 2012).

30 See e.g. Lena Maria Schaffer and Thomas Bernauer, 'Explaining Government Choices for Promoting Renewable Energy' (2014) 68 *Energy Policy* 15; Nicholas Apergis and James E Payne, 'Renewable Energy Consumption in Economic Growth: Evidence from a Panel of OECD Countries' (2010) 38 *Energy Policy* 656; Reinhard Haas et al., 'Promoting Electricity from Renewable Energy Sources – Lessons Learned from the EU, US and Japan' in Fereidoon P Siosanshi (ed.), *Competitive Electricity Markets: Design, Implementation, Performance* (Elsevier Science, 2008) 419; Katrin Jordan-Korte, *Government Promotion of Renewable Energy Technologies: Policy Approaches and Market Development in Germany, the United States, and Japan* (Gabler Research, 2011).

31 See e.g. Kat Cheung and IEA, '*Integration of Renewables – Status and Challenges in China*' (Working Paper, OECD, 2011); Cui Huang et al., 'Government Funded Renewable Energy Innovation in China' (2012) 51 *Energy Policy* 121.

32 Directive 2009/28/EC of the European Parliament and of the Council of 23 April 2009 on the promotion of the use of energy from renewable sources and amending and subsequently repealing Directives 2001/77/EC and 2003/30/EC [1993] OJ L 140/16.

and the IRENA Statute.[33] This analysis is used to develop a comprehensive scholarly understanding of how different countries legislatively define renewable energy, what they are trying to achieve through the adoption of these laws and which regulatory support mechanisms they utilise. Based on this understanding of the different conceptions of renewable energy and the rationales for countries legislating in the sector, an assessment is then made about the likely future development of national renewable energy laws. Will countries actively seek to engage in regulatory competition through their national renewable energy laws in order to attract investment and fulfil other industrial policy objectives such as developing a foothold in the renewable technology export market? Or will they naturally diverge to reflect the local preferences of the country's citizenry, as well as its unique energy security, political, economic, social, legal and environmental contexts? Will more countries seek to adopt the preferred method among the European Union of a convergence of national laws through cooperation and coordination processes? Or, with the increasing commercialisation of renewable energy technologies throughout the world, will countries consider harmonising or unifying their renewable energy laws to reduce the transaction costs borne by international companies? Without understanding the different countries' approaches to their national renewable energy law, the extent and scale of the patterns and tensions that are emerging globally between the differing approaches are going unnoticed. This book seeks to remedy this oversight by providing the first comprehensive scholarly analysis of the national renewable energy laws of all countries that have such laws.

1.2 HYPOTHESIS AND RESEARCH METHODOLOGY

The hypothesis tested in this research is that, as renewable energy sources and technologies used around the world become commercialised and more widely adopted, renewable energy laws will also come under pressure to harmonise or converge to facilitate trade, improve information sharing and ease administration. This will see similar legislative definitions for renewable energy being adopted, similar legislative objectives found in national renewable energy laws and regulatory support mechanisms acquiring similar designs and being adopted in similar combinations by different countries throughout the world.

33 *Statute of the International Renewable Energy Agency*, opened for signature 26 January 2009, [2009] ANTIF 23 (entered into force 8 July 2010).

In order to test this hypothesis, the questions addressed in this research are:

1. Which countries have a national framework law to govern or promote the accelerated deployment of renewable energy? (List of National Renewable Energy Laws)
2. Which energy sources are recognised as renewable energy sources within the legislative definitions in the national renewable energy laws of different countries? (Chapter 2)
3. What is the theoretical rationale for governments legislating to support the accelerated deployment of renewable energy? (Chapter 3)
4. What are the stated legislative objectives for supporting the accelerated deployment of renewable energy in the primary legislation? (Chapter 4)
5. How have regulatory support mechanisms been designed to accelerate the deployment of renewable energy in different countries? (Chapter 5)
6. Given the benefits of national renewable energy laws becoming more similar, how will regulatory support mechanisms likely develop in the future? Will they be unified, harmonised, converge, diverge or actively compete through regulatory competition? (Chapter 6)

This research involved the study of the primary piece of renewable electricity legislation in each of the 113 countries that had a national renewable energy law on 1 August 2018. This original research was aided by the adoption of a comparative mixed methodological approach, using the available quantitative data and qualitative research methods such as statutory interpretation, comparative analysis, and doctrinal research, supplemented by analysis of primary and secondary legal, political, economic and historical sources. The breadth of research was designed to overcome problems of source coverage and source bias and offer a deeper and more nuanced assessment of the regulatory regime.

The comparative mixed methodological approach best facilitated a study of the similarities and differences of the primary legislative instruments covering the renewable energy sector. The comparative analysis was based on both a textual and functional analysis of the national renewable energy laws drawing upon the research of eminent scholars in these fields.[34] This approach was

[34] This research drew upon the methodological approaches of Alan Watson, 'Comparative Law and Legal Change' (1978) 37 *Cambridge Law Journal* 313, 317; John Henry Merryman, 'Comparative Law Scholarship' (1998) 21 *Hastings International and Comparative Law Review* 771, 775; John Bell, 'Legal Research and the Distinctiveness of Comparative Law' in Mark Van Hoecke (ed.), *Methodologies of Legal Research: Which Kind of Method for Which Kind of Discipline?* (Hart, 2011) 155, 158; Konrad Zweigert and Hein Kotz, 'The Method of Comparative Law' in *An Introduction to Comparative Law* (3rd edn, Clarendon Press, 1998) 34.

adopted because 'even the same legal rules and institutions have widely different consequences, depending on the operation of "contextual" factors'.[35]

One of the most difficult aspects of this research was identifying and locating the national renewable energy laws in every country that possessed them, as well as confirming that such laws did not exist in the remaining countries. While the renewable energy laws were generally publicly available, they were not compiled in a single database. Thus, the first step in this process was to identify the correct databases to search for the relevant laws for each country. Where a regularly updated and authoritative national collection of legislation could be identified for a country, this was the preferred source of the legislation.[36] Once the correct collection of legislation was identified, searches were then carried out. English was used as the language of the search where this was an available option. Where this was not available, online and hard copy foreign language dictionaries were used to identify the correct search terms in the official language of the host government. This meant that searches in Catalan were used for the Andorran legislative database, Dari for the Afghan legislative database and Arabic for the Yemeni legislative database. The searches used a wide range of possible terms to capture the subject matter of the various laws that may be functionally equivalent: 'renewable energy', 'renewable energy sources', 'renewable electricity', 'renewable energy resources', 'energy from renewable sources', 'new and renewable energy', 'renewable and sustainable energy sources', 'alternative energy', 'green power', 'non-conventional energy', 'non-traditional energy sources' and 'non-fossil energy'. The search parameters excluded legislation that:

- was not national or supranational (thereby excluding consideration of state, regional or provincial laws such as the Ontario Feed-in Tariff); or
- had ended, been superseded or planned but was not in force on 1 August 2018; or
- was specifically focused on heating and cooling or transport rather than electricity.

Out of the total 113 countries with national renewable energy laws, there were laws studied that were drafted in 47 different languages. Preference was given to authoritative versions of the renewable energy laws where these were available in English. Where no authoritative version of the law was available

35 David Nelken and Johannes Feest (eds.), *Adapting Legal Cultures* (Hart Publishing, 2001) 9.

36 The numerous sectoral databases of policies and measures were not used as a reliable source of the laws, as they were often found to be incomplete or out of date. They did however play a valuable role in cross-checking the data collected.

in English, an official or non-binding translation of the authoritative version conducted by or for a governmental body or organisation was then used. If no official translation existed, the laws then had to be legally translated from their original language into English.[37]

Once the relevant laws were identified, located and where necessary, translated, they were then coded to make sense of the data collected. The coding process focused on two separate sections of the legislation (or their functional equivalent): the legislative definition of renewable energy (or the synonym adopted by the jurisdiction) and the legislative objectives. The coding process was important as the codes were used to examine the similarities, differences and frequency of key concepts and themes between the national renewable energy laws of different countries. They were also used to look at the sequencing of concepts and themes (i.e. were concepts and themes mentioned in a certain order? Was this indicative of the relative priority assigned to that concept or theme in the legislation?).

This methodology was selected because it enabled a wider geographic scope and thus broadened the applicability of the research by providing a deeper and more nuanced understanding of the outcomes of different regulatory approaches. The use of a large n-sample meant that generalisable inferences could be drawn and conclusions reached about the current state of renewable energy law globally. This provided a better understanding of the emerging norms for renewable energy laws against which the laws of individual countries can now be compared, as well as furthering knowledge of the way that specific regimes fit the particular developmental or other characteristics of the countries studied.

1.2.1 *Limitations of the Research Methodology Adopted*

At this juncture, several limitations to the research methodology adopted should be acknowledged. First, due to the number and diversity of the countries studied in this research, it was not possible to track the changes in the law over time. Nor was it possible to consider the potential layering impact

[37] The translation of these laws was conducted by legally trained translators, predominantly from a professional legal translation service, Linguistico. In addition, some senior year law students from the Faculty of Law at the University of Sydney, who were native speakers in the original language, also volunteered to assist with this exercise. The names of these students are Stephanie Watson, Tallulah Bur and Ashley Richards (French), Ellen Marie O'Brien (Indonesian), Levi Romanov (Russian) and Laura Peck and Mitchell Cleaver (German).

of regional, provincial, state or local government laws operating in addition to the national law. This limited the ability at this point in the research to be able to make a thorough assessment of whether the laws are currently converging or diverging, although in the latter chapters, a number of hypotheses are drawn as to the likely future development of the laws based on the existence of similarities and differences between them.

Second, the range of languages in which the primary laws were drafted created a language barrier. In order to overcome this barrier, it would have been preferable to have had the entirety of all of the relevant laws translated into English using professional legal translators. This would have enabled more accurate statutory interpretation that considered the section in the context of the legislation as a whole. However, with the laws of 113 countries being studied, it was simply not feasible.

Despite these limitations, when the primary data were combined with the substantial secondary resources used throughout the research, it enabled significant and methodologically robust results and conclusions to be reached.

1.3 STRUCTURE OF THE BOOK

This book comprises seven chapters divided into three parts. This first chapter explains the background to the book, along with the problem being addressed and the significance of this research as the first academic study of the primary renewable energy legislation in each of the 113 countries that have a national renewable energy law. It also briefly outlines the methodological approach adopted.

1.3.1 *Part I What Is Renewable Energy? A Case of Conceptual Consensus*

The first substantive part of the book, comprising Chapter 2, seeks to discover whether a common understanding of the concept of 'renewable energy' has developed in the laws of countries seeking to accelerate its deployment. In order to do this, the subject matter of national renewable energy laws is examined to assess the nature of and levels of support for the various energy sources and renewable energy technologies identified within the legislative definitions of renewable energy. In doing so, it provides the first global snapshot of the frequency of renewable energy sources being included in renewable energy laws. It also broadly explores how renewable energy is generated from those renewable energy sources, and highlights the current debates around whether certain energy sources should be considered to be renewable.

1.3.2 *Part II Why Do Countries Intervene in the Renewable Energy Sector? A Case of Normative Divergence*

Once the subject matter of the national renewable energy laws is understood, the second part of the book focuses on the rationale of countries engaging in regulatory intervention to support the accelerated deployment of renewable energy. It examines the justifications derived from economic theory for regulatory intervention in the renewable energy sector (Chapter 3) and then compares this to the legislative objectives contained in the national renewable energy laws of countries who have legislated in this area (Chapter 4). In this part, the research focuses on normative consensus or divergence, that is, whether governments are intervening in the renewable energy sector for the same or different reasons.

Chapter 3 starts from the premise that from an economic perspective, well-functioning, competitive global and domestic markets generally provide the most suitable basis for optimal and timely investment decision-making. It then considers how the features of electricity generally, and generation from renewable sources more specifically, warrant special regulatory treatment due to it being a 'mixed good'. This chapter then examines the market failures within the energy sector and how they negatively impact on the deployment of renewable energy, including: (i) the presence of unpriced negative and positive externalities in the energy sector; (ii) spillovers and learning effects; and (iii) information asymmetries. A range of market barriers that affect the renewable energy sector in different countries are also examined. This forms the basis of an inquiry into whether government intervention in the renewable energy sector is warranted from an economic perspective. Further consideration is given to how such an intervention should be structured to minimise distortions to the market.

Chapter 4 explores the role of legislative objectives in the renewable energy laws of countries to assess why countries are regulating to promote the deployment of renewable energy. In particular, the research in this chapter seeks to better understand whether countries are regulating merely to try to correct the market failures that exist within the sector identified in Chapter 3, or whether countries are trying to achieve something more through their renewable energy laws. Chapter 4 begins by evaluating the role of legislative objectives before critically reviewing the previous research on the reasons why countries seek to accelerate the deployment of renewable energy. This allows an assessment to be made about whether existing understandings of the rationale for government interventions within the renewable energy sector are correct and generalisable across all the countries with national renewable energy laws.

The focus of the chapter then shifts to each of the twenty-eight different categories of legislative objectives found within the national laws, before examining the issue of conflicting legislative objectives. The chapter concludes with some recommendations of how legislative objectives in renewable energy laws may be improved in the future.

1.3.3 *Part III What Role Do Regulatory Support Mechanisms Play in National Renewable Energy Laws? A Case of Substantive Divergence*

The third part of this book examines the range of regulatory support mechanisms used by countries when they intervene in the markets to support the accelerated deployment of renewable energy (Chapter 5), before considering whether these regulatory support mechanisms are likely to converge or diverge over time (Chapter 6).

Chapter 5 highlights the range of regulatory support mechanisms available to countries seeking to ameliorate the market failures and barriers impeding their ability to achieve their legislative objectives. It considers the criteria that countries use to select appropriate design options, common classification systems for the mechanisms, as well as providing a detailed account of the common characteristics of the different types of regulatory support mechanisms used within the renewable energy sector. In particular, feed-in tariffs, feed-in premiums, renewable portfolio standards/quota obligations, green certificate trading/renewable energy credits, competitive tendering (auction bidding), subsidies, net metering, loans, rebates, investment tax credits, production tax credits, green power schemes, grants, research and development support and other indirect mechanisms are examined. This chapter further considers the implications of the growing use of combinations of multiple regulatory support mechanisms within the sector. The chapter concludes with a brief discussion of some of the lessons learnt for the successful design and implementation of regulatory support mechanisms within the renewable energy sector.

Chapter 6 considers the likely future development of regulatory support mechanisms in the national renewable energy laws. This chapter draws together the analysis conducted in previous chapters to examine whether regulatory support mechanisms found in national renewable energy laws are likely to internationally harmonise or converge in the future as the renewable energy technologies become more similar, widespread and bankable. Or are the regulatory support mechanisms more likely to diverge internationally or even actively compete via regulatory competition as countries continue to pursue their national self-interest? This chapter reviews why it is desirable for national renewable energy laws to become more similar internationally, despite the fact that this will mean a loss of local preferences. This chapter further examines the development of the European Union (EU) renewable energy law, as well as some of the international trade conflicts that have emerged in the past ten years within the renewable energy sector in order to understand how international legal institutions are shaping the development of regulatory support mechanisms within national renewable energy laws.

1.3.4 *Conclusion*

The book concludes in Chapter 7 by returning to the research questions posed in Chapter 1 to test the hypothesis that, as different techniques for generating renewable energy become commercialised and the manufacturing of renewable technologies becomes more concentrated in particular countries, renewable energy laws will also come under pressure to harmonise to facilitate trade, improve information sharing and ease administration. To test this hypothesis, Part I of this research focused on the fundamental concepts of how renewable energy is defined in law and which energy sources are recognised as renewable by different countries. The purpose of this Part was to test whether countries were legislating on the same subject matter to see if a degree of conceptual consensus was emerging between the renewable energy laws of different countries. As elucidated in Chapter 2, there was a striking degree of conceptual consensus in the energy sources recognised as 'renewable energy' within the legislative definitions of different countries, especially for the most commercialised renewable energy sources and technologies. This showed that, despite the different approaches to the form and content of the legislative definitions, conceptual consensus was already evident between many countries' national renewable energy laws. This conceptual consensus may provide the basis for later international harmonisation or legislative convergence in years to come.

In Part II of the book, the emphasis shifted to considering whether different countries shared a common rationale for regulatory intervention in the renewable energy sector. This was testing whether normative consensus was present in the renewable energy laws of different countries. Chapter 3 showed that economic theory operates on the premise that governments should only intervene in the market to correct the three market failures present in the renewable energy sector: the presence of unpriced negative and positive externalities in the energy sector; spillovers and learning effects; and information asymmetries. This suggested that there should be normative consensus between different countries' laws. However, when these market failures were compared to the legislative objectives contained in the national renewable energy laws considered in Chapter 4, this was found not to be the case. Rather, a much broader range of legislative objectives were identified than in any of the previous research, with countries having legislative objectives that fell within eight themes and twenty-eight different subcategories. Further, while some legislative objectives that addressed the market failures addressing unpriced externalities such as 'energy security' and 'diversify supply' were highly weighted in priority, others that also addressed the issue of unpriced environmental externalities such as 'reduce greenhouse gas emissions and address climate change' were lowly weighted in priority. In this part, it was evident that countries continue to engage in regulatory intervention not merely to address the market failures that exist in the sector but rather to address a much broader range of, often domestic, concerns. As a result, while the economic theory suggests that normative consensus between different countries' rationales for regulatory intervention in the renewable energy markets should exist, it was clear from Chapter 4 that a significant degree of normative divergence was present between the laws.

In Part III, the research considered whether there was substantive convergence between the regulatory support mechanisms adopted by different countries in their national renewable energy laws. It further considered the likely future development of the national renewable energy laws. In Chapter 5, it was established that while common types of regulatory support mechanisms exist, such as feed-in tariffs or feed-in premiums, renewable portfolio standards or competitive tendering, every country has different design features and adopts different combinations of them to accelerate their deployment of renewable energy. This suggests that no one mechanism or combination of mechanisms will meet the needs of every country. Thus, the starting position for most countries seems to be one of substantive divergence, with different regulatory support mechanisms being designed and implemented in different countries. This view was reinforced by the case studies in Chapter 6, which showed that a number of countries are pressing the divergence one step further and are actively adopting the process of regulatory competition through their laws.

The initial hypothesis of this research – that over time the national renewable energy laws should come under pressure to harmonise as countries seek to engage in either a 'race to the bottom' or a 'race to the top' – was proven correct at least initially in the European Union (EU), which had two separate failed harmonisation attempts. However, the EU Member States have successfully resisted this pressure, with the Member States explicitly inserting a reservation into the Treaty of the Functioning of the European Union that will ensure their ability to control their own energy mix and support systems going forward. This has led to the EU having to, at least temporarily, drop its expressed preference for a 'top-down' harmonisation under the Renewable Energy Directive in favour of adopting a process of cooperation and coordination. That said, the EU may now be seeking to use its exclusive competence in competition law to achieve harmonisation of the regulatory support mechanisms by stealth under the State Aid Guidelines. Meanwhile some other countries around the world, such as China and Japan, have proactively pursued a concerted policy of regulatory competition as they seek to become the lead market. Thus, contrary to the initial hypothesis, this research demonstrates that while there is strong conceptual consensus within the legislative definitions of renewable energy, significant normative and substantive differences still exist within the national laws governing the promotion or accelerated deployment of renewable energy of many countries.

PART I

WHAT IS RENEWABLE ENERGY?
A CASE OF CONCEPTUAL CONSENSUS

2

The Renewable Energy Sources Used for Electricity Generation

Formulating a legislative definition of 'renewable energy', or its functional equivalent, is important in countries that have a primary legislative framework governing or promoting renewable energy law. The presence of a legislative definition delimits the core subject matter of such laws and provides greater certainty to the sector about which energy sources and technologies should be supported. The primary classification of energy, which has been adopted almost universally, is the dichotomous division of energy fuel sources between fossil fuels and renewable sources. This division is often reflected in the definition of renewable energy adopted by different jurisdictions. However, because the definition of 'renewable energy' is often uncertain and over-inclusive, debate has raged as to whether certain types of energy should be deemed renewable and thus benefit from the programmes designed to accelerate their adoption, or whether they should be excluded on other environmental and policy grounds.

In particular, concerns have been raised about the environmental impacts of some arguably renewable technologies such as large-scale hydropower, woody biomass and peat, leading to arguments that they are not sustainable and thus should not be captured within the definition of 'renewable energy'. Conversely, others have argued that some technologies that are dependent on energy sources, which are arguably not renewable, such as geothermal and nuclear energy, should be included in the definition of renewable energy if the purpose of the legislation is to encourage the accelerated deployment of low carbon electricity generation. It is important to note that 'no form of energy is free from monetary and environmental costs',[1] thus the final determination of what constitutes renewable energy is often the result of a politically mediated

1 Richard L Ottinger and Adrian J Bradbrook (eds.), *UNEP Handbook for Drafting Laws on Energy Efficiency and Renewable Energy Resources* (UNEP/ Earth Print Limited, 2007) 2.

debate. This means that other factors such as the impact on local employment prospects and protecting indigenous energy sources for energy security reasons also tend to figure in debates about whether to include a particular energy source as 'renewable'.

This chapter considers the different sources of renewable energy and renewable energy technologies that are accepted by countries within their renewable energy laws. In doing so, it draws upon the results of statutory interpretation of the legislative definitions of renewable energy to provide the first global snapshot of the frequency of renewable energy sources and technologies being included in renewable energy laws. This chapter also broadly explores how electricity is generated from those renewable energy sources, and highlights the current debates around whether certain energy sources should be considered to be renewable. For the remainder of the chapter, each potential source of renewable energy will be considered in the order of its level of acceptance in the legislative definitions of the different countries' national renewable energy laws. Where the inclusion of a particular energy source as renewable is the subject of debate, this will also be examined.

2.1 WIND ENERGY

Wind energy is the second fastest growing form of renewable energy,[2] with onshore wind being one of the most cost-competitive renewable energy technologies.[3] Globally, there was 539 gigawatt (GW) of installed wind power capacity by the end of 2017.[4] As a result, wind energy is the most commonly recognised renewable energy source within the legislative definitions of renewable energy, being recognised by 108 out of 113 countries with renewable energy laws. Wind energy is also included in both the EU and the IRENA definitions of renewable energy. It should be noted that Greece only includes onshore wind within their legislative definition. In a further three countries, it is unclear whether wind energy is included or excluded from the definition of renewable energy due to the presence of ambiguous definitions or a failure to define renewable energy. The only country explicitly to exclude all forms of wind energy from their legislative definition is Malaysia. This may reflect the historical, though some claim inaccurate, view that Malaysia

2 IEA, *Renewables Information 2018* (OECD/IEA, 2018) xiv.

3 IRENA, *Renewable Power Generation Costs in 2017* (IRENA, 2017) 15.

4 Global Wind Energy Council, Global Wind Report: Annual Market Update 2017 (GWEC 2018), 16.

possessed a poor wind resource and that their efforts were better directed at alternative energy sources.[5]

Wind energy is derived from the uneven solar heating and cooling of the Earth's surface and atmosphere.[6] These uneven patterns of heating and cooling create currents of air that can be exploited to unlock their energy potential.[7] Wind energy is produced when kinetic energy, in the form of wind, passes through the blades of the wind turbine and turns a rotor connected to an electrical generator.[8] This process transforms the kinetic energy into electrical energy.[9] The two main types of wind energy project currently used to generate electricity are onshore wind energy and offshore wind energy.

While the sizes of wind turbines vary with the size of the generator, diameter of the blades, hub height and conditions specified in the planning permission, a standard total height[10] for an onshore commercial wind turbine is approximately 140m.[11] The number of turbines installed in an individual wind energy project will depend on the location and capacity of the turbines, but may vary from a single turbine up to several hundred.[12] While the technology used for onshore and offshore projects is similar, the scale of the turbines used in offshore projects tends to be much greater than those used onshore. This is because of the greater wind speeds that tend to be generated offshore, warranting the use of larger turbines that may be more robust. Due to the greater size of these offshore turbines, they commonly require foundations to be drilled into the ocean floor, and then underwater transmission lines to be installed, with an onshore grid connection.

Recent innovations in the offshore wind sector include the development of floating platforms for wind turbines, such as those used in the 30MW

5 Lip-Wah Ho, 'Wind Energy in Malaysia: Past, Present and Future' (2016) 53 *Renewable and Sustainable Energy Reviews* 279–95.

6 Jeremy Firestone and Jeffrey P Kehne, 'Wind' in Michael B Gerrard (ed.), *The Law of Clean Energy: Efficiency and Renewables* (American Bar Association, Section of Environment, Energy and Resources, 2012) 361; Jefferson W Tester et al., *Sustainable Energy: Choosing Among Options* (The MIT Press, 2005) 617.

7 Tester et al., above n 6, 617.

8 See generally IRENA, *Wind Energy* (2018) <www.irena.org/wind> (accessed 17 September 2018).

9 Tester et al., above n 6, 409.

10 Total height is calculated by adding the blade height to the tower height of the wind installation.

11 See e.g. Ryan Wiser and Mark Bolinger, 2016 Wind Technologies Market Report, Office of Energy Efficiency and Renewable Energy, US Department of Energy, Chapter 4. (2016) <www.energy.gov/sites/prod/files/2017/10/f37/2016_Wind_Technologies_Market_Report_101317.pdf>.

12 Alexandra Wawryk, 'Solar and Wind Energy' in Ottinger and Bradbrook (eds.), above n 1, 161.

Hywind Scotland Project developed by Equinor.[13] This is an important development for the sector, as approximately 80 per cent of the world's offshore wind potential is located in deep waters (greater than 60m), which are unsuitable for fixed foundations.[14] This could be an attractive option for countries, such as Japan, which have limited suitable locations for offshore wind in shallow coastal waters. While floating wind turbines are not currently commercially cost-competitive, it is likely that prices will rapidly decline as the technology is adopted, with some estimates predicting that the use of floating platforms could be cost-competitive by the late 2020s.[15]

Both onshore and offshore wind projects allegedly possess a number of negative environmental impacts. Commonly cited problems include the visual impact on the landscape,[16] the impact on bird and bat life when turbines are improperly located in their flight paths,[17] the impact of drilling foundations for offshore turbines on cetaceans and the marine environment[18] and noise emissions.[19] Complaints have also been made about electromagnetic interference with television, radio and radar signals.[20]

13 See e.g. IRENA, *Floating Foundations: A Game Changer for Offshore Wind Power* (IRENA, 2016) <www.irena.org/-/media/Files/IRENA/Agency/Publication/2016/IRENA_Offshore_Wind_Floating_Foundations_2016.pdf>; European Wind Energy Association, Deep Water: The Next Step for Offshore Wind Energy, July 2013 <www.ewea.org/fileadmin/files/library/publications/reports/Deep_Water.pdf>.

14 Equinor, *Hywind Offshore Wind Project* (2018) <www.equinor.com/en/what-we-do/hywind-where-the-wind-takes-us.html#floating-wind>.

15 Ryan Wiser and Mark Bolinger, 2016 Wind Technologies Market Report, Office of Energy Efficiency and Renewable Energy, US Department of Energy, 62 <www.energy.gov/sites/prod/files/2017/10/f37/2016_Wind_Technologies_Market_Report_101317.pdf>.

16 See e.g. ERM and REARK Research, *Study into Community Attitudes About Wind Farms in the NSW Southern Tablelands* (Epuron, 2007) 1; Wawryk, above n 12, 166; Ernest E Smith, 'Wind Energy: Siting Controversies and Rights in Wind' (2007) 1 *Environmental Energy Law & Policy Journal* 281, 283; *Rankin v. FPL Energy LLC*, 266 SW 3d 506, 509-11 (Tex Ct App, 2008).

17 Smith, above n 16.

18 Tester et al., above n 6, 636.

19 See e.g. the following court cases from the United States alleging that noise from neighbouring wind farms constituted a substantial and unreasonable interference with the property rights of the plaintiff: *Rose v. Chaikin* 187 NJ Super 210, 214–20 (Ch Div, 1982); *Rassier v. Houim*, 488 NW 2d 635, 638 (ND, 1992); *Burch v. Nedpower Mount Storm LLC*, 647 SE 2d 879, 885, 891 (WV, 2007); *Rankin v. FPL Energy LLC*, 266 SW 3d 506, 513 (Tex Ct App, 2008); *O'Dell v. FPL Energy LLC*, No. 06-502, slip op (235 JD Tex, 2006) (settled prior to judgment); *Bruck v. Gamesa Wind US LLC*, No. 06-0129, slip op (271 JD Tex, 2006) (concluded in a nonsuit); *Porter v. Gentry County Commission*, No. 08-6029-CV-SJ-FJG, slip op (DC Mo, 2008).

20 Tester et al., above n 6, 637.

In relation to noise emissions, some opponents of wind energy projects claim that the presence of low frequency noise and infrasound has caused some individuals to develop 'Wind Turbine Syndrome'.[21] The alleged symptoms of Wind Turbine Syndrome include direct physiological effects such as heart attacks, vertigo and tinnitus.[22] However, reports from both the National Health and Medical Research Council[23] in Australia and an expert review panel[24] commissioned by the American Wind Energy Association and the Canadian Wind Energy Association concluded that 'there [was] no published scientific evidence to support adverse effects of wind turbines on health'.[25] Indeed, even the more politically driven Australian Commonwealth Senate Committee inquiry into 'The Social and Economic Impacts of Rural Wind Farms' stated that 'adverse health effects may be caused by wind turbines but they may be caused by factors other than noise and vibration, such as stress related to sleeplessness or perceptions of harm'.[26] This conclusion has been echoed by other studies, with Chapman arguing that for people who are already predisposed to anxiety disorders, the placement of a wind turbine on a neighbouring property may be a source of additional stress.[27] In addition, Wind Turbine Syndrome has not been recognised by any international disease classification database.[28]

In respect of the other negative environmental impacts, many of these have diminished as issues in recent years, due to technological developments and greater attention being paid in the planning process to the siting and potential impacts of wind farms.[29]

21 Nina Pierpont, *Wind Turbine Syndrome: A Natural Experiment* (K-Selected Books, 2009).

22 Jane Davis, *General Statement by Jane & Julian Davis* (16 April 2009) European Platform Against Windfarms <www.epaw.org/victims.php?lang=en&article=t1>.

23 NHMRC, *Wind Turbines and Health* (July 2010) NHMRC <www.nhmrc.gov.au/_files_nhmrc/file/publications/synopses/public_statement_wind_turbines_and_health>.

24 W David Colby et al., *Wind Turbine Sound and Health Effects: An Expert Panel Review* (AWEA/CWA, 2009) <www.canwea.ca/pdf/talkwind/Wind_Turbine_Sound_and_Health_Effects.pdf>.

25 NHMRC, above n 23, 1.

26 Community Affairs Committee, Senate of the Parliament of Australia, *The Social and Economic Impact of Rural Wind Farms* (2011) chapter 2 [2.99].

27 See e.g. Simon Chapman and Fiona Crichton, *Wind Turbine Syndrome: A Communicated Disease* (Sydney University Press, 2017); Simon Chapman, 'Wind Farms and Health: Who Is Fomenting Community Anxieties?' (2011) 195 *Medical Journal of Australia* 495; see also Loren D Knopper and Christopher A Ollson, 'Health Effects and Wind Turbines: A Review of the Literature' (2011) 10 *Environmental Health* 78.

28 Simon Chapman, 'Much Angst over Wind Turbines Is Just Hot Air', *Sydney Morning Herald* (online), 21 December 2011 <www.smh.com.au/federal-politics/political-opinion/much-angst-over-wind-turbines-is-just-hot-air-20111220-1p3sb.html>.

29 Firestone and Kehne, above n 6, 369, 371–2.

2.2 SOLAR ENERGY

Solar radiation is the largest and most accessible renewable resource on Earth. It is more evenly distributed than other forms of renewable energy, which may require access to a water source such as hydropower or tidal or wave power, access to wind such as wind power or access to organic waste such as biomass.

Solar energy has two primary purposes: to generate electricity and to provide heat. There are two main methods by which solar energy may be used to generate electricity: photovoltaic solar power and concentrated solar thermal power.

2.2.1 *Photovoltaic Solar*

In terms of installed capacity, photovoltaic solar generation is the most widely used form of solar technology with 402GW installed globally by the end of 2017.[30] It is also the most commonly supported solar technology within the legislative definitions of renewable energy, being recognised by 107 countries, as well as the EU and the IRENA. Only Brazil[31] and Finland clearly do not promote renewable electricity generation from photovoltaic solar, while another four countries lack clarity on whether photovoltaic solar is included or excluded by their definitions.

Photovoltaic technology works by converting solar photons from sunlight into direct electrical current through the process of freeing electrons from their atomic bonds as they travel across semiconductor materials.[32]

Photovoltaic solar energy is the fastest growing source of renewable energy capacity, with the growth in global capacity averaging 32 per cent annually between 2013 and 2018.[33] It also offers a number of advantages when compared to other technologies. Photovoltaic solar cells are relatively easy to install,[34] scalable from small to large commercial applications,[35] and due to the lack of

[30] REN21 Secretariat, 'Renewables 2018 Global Status Report' (Report, Renewable Energy Policy Network for the 21st Century, 2018) 19.

[31] Note that while Brazil does not support PV solar under its national renewable energy legislation, it does provide other support for net metering and community solar under subsidiary regulations.

[32] IRENA, *Solar Photovoltaics* (IRENA, 2012) 4.

[33] REN21 Secretariat, above n 30, Figure 24.

[34] Kevin Bullis, 'Solar Panels That Configure Themselves', *MIT Technology Review* (online), 20 November 2014 <www.technologyreview.com/news/532731/solar-panels-that-configure-themselves/>.

[35] Tester et al., above n 6, 572; Stewart Needham, 'Renewable Energy Technologies Update' (Background Note, Parliament of Australia, 2009) 7; Craig M Kline, 'Solar' in Michael B Gerrard (ed.), *The Law of Clean Energy: Efficiency and Renewables* (American Bar Association Section of Environment, Energy and Resources, 2012) 392.

moving parts (unless the cells have sun-tracking hardware installed), have low maintenance costs.[36] They also offer flexibility,[37] with photovoltaic solar cells capable of being directly integrated into both consumer goods and buildings, and operated either in static or mobile installations.[38] Other advantages of photovoltaic solar technologies include environmental benefits such as the lack of greenhouse gas emissions produced during generation[39] and that the technologies operate noise-free.[40] Photovoltaic solar cells have an expected lifespan of between ten and thirty years,[41] though other components of the modules such as the inverters and batteries need to be replaced every five to ten years.[42]

As with all energy technologies, photovoltaic solar is not without its downsides. First, a comparatively large amount of electricity is used in the manufacture of photovoltaic solar technology relative to some other renewable technologies, with the effective energy payback time being between two and five years of domestic operation[43] and between one and three years of commercial operation.[44] Further, as the electricity used in solar cell production is generally grid supplied, in countries that do not have 100 per cent renewable generation, the electricity produced using photovoltaic solar is said to 'inherit' the greenhouse gas (GHG) emissions created during the manufacture of the technology.[45] However, even when these emissions are factored in for photovoltaic solar technologies, the emissions are still several times lower than for coal-based generation.[46] Second, where the solar modules are part of a large-scale ground-based project, rather than being attached to buildings, they can take up

36 Ibid.

37 Tester et al., above n 6, 582; IEA, *Renewable Energy: RD&D Priorities* (OECD/IEA, 2006), 109; IEA, *Technology Roadmap: Solar Photovoltaic Energy* (OECD/IEA 2014) 5.

38 Ibid; see also IEA, *Transition to Sustainable Buildings: Strategies and Opportunities to 2050* (OECD/IEA, 2013) 139.

39 Ibid.

40 Tester et al., above n 6, 572.

41 Tester et al., above n 6, 576; Kline, above n 35, 393; IEA, *Technology Roadmap: Solar Photovoltaic Energy*, above n 37, 12.

42 IRENA, *Boosting Solar PV Markets: The Role of Quality Infrastructure* (IRENA, 2017) 29; the anticipated lifespan of these components of the solar modules is even lower in hot climates.

43 Tester et al., above n 6, 582. See also Alison Potter, *Solar Panel Payback Time* (10 July 2018) <www.choice.com.au/home-improvement/energy-saving/solar/articles/solar-panel-payback-times>.

44 IEA, *Technology Roadmap: Solar Photovoltaic Energy*, above n 37, 31.

45 IEA, *Renewable Energy: RD&D Priorities*, above n 37, 117.

46 Ibid.

a significant area of land.[47] This may lead to potential land use conflicts in the future around urban centres, as due to the load losses experienced when electricity is transported long distances, the generators need to be positioned close to the communities that they are powering. Finally, concerns have also been expressed about the use of toxic and flammable/explosive gases and toxic metals, such as cadmium, in the manufacture of the solar cells.[48] A particular concern is the disposal of the cells at the end of their life, leading to current research and development to facilitate the recycling of solar cells and to decrease the amounts of these materials required in future solar cells.[49]

2.2.2 *Concentrated Solar Thermal Technology*

Concentrated solar thermal energy attracts similar levels of support within the legislative definitions of renewable energy as photovoltaic solar energy, with 106 countries promoting it, as well as the EU and the IRENA. The two countries that do not support photovoltaic solar energy in their primary legislation, Brazil and Finland also do not support concentrated solar thermal energy, and nor does Malaysia. In four countries it is not apparent if concentrated solar thermal energy is included or excluded from their legislative definition.

Concentrated solar thermal technologies work by capturing and concentrating solar radiation onto a receiver, where it is converted into heat.[50] The heat created is then either used directly or used to boil a working fluid to produce steam to power the turbines.[51] Several different power conversion systems may be used in this process, including Rankine, Brayton, combined or Stirling cycles.[52] There are three main categories of concentrated solar thermal technologies: parabolic troughs, parabolic dishes and solar towers.[53] These technologies all use a similar process despite using different shaped receivers/concentrators.

47 IEA, *Renewable Energy: RD&D Priorities*, above n 37, 118; Tester et al., above n 6, 583; see also IEA, *Solar Energy Perspectives* (OECD/IEA, 2011) 208–9.

48 IEA, *Solar Energy Perspectives*, above n 47, 115; Tester et al., above n 6, 580, 583.

49 IEA, *Solar Energy Perspectives*, above n 47, 115.

50 Tester et al., above n 6, 562; IRENA, *Concentrating Solar Power* (IRENA, 2012) 4.

51 Tester et al., above n 6, 562; IRENA, *Concentrating Solar Power*, above n 50, 4; Needham, above n 35, 1.

52 IEA, *Renewable Energy: RD&D Priorities*, above n 37, 153.

53 Tester et al., above n 6, 562–9; IEA, *Technology Roadmap: Solar Thermal Electricity* (OECD/IEA, 2014) 11.

Concentrated solar thermal technologies offer a number of advantages. They can be sized for small scale (10kW) or grid-connected applications (up to 377MW);[54] they can provide energy storage capabilities;[55] and they can be integrated into conventional thermal plants to provide reliable dispatchable power.[56] The IRENA has predicted that large-scale concentrated solar thermal power will be cost-competitive in the next ten years, based on assumptions about cost reductions in the technology, potential price increases for fossil fuels and the internalisation of other costs such as carbon emissions.[57]

2.3 BIOMASS

Biomass is defined by the OECD as 'any plant matter used directly as fuel or converted into other forms before combustion. Included are wood, vegetal waste (including wood waste and crops used for energy production), animal materials/wastes, sulphite lyes, also known as "black liquor" (an alkaline-spent liquor from digesters in sulphate production or soda pulp during the manufacture of paper where the energy content derives from the lignin removed from the wood pulp) and other solid biomass'.[58] Other definitions include all organic matter, including human and animal wastes and marine life (particularly algae).[59]

Electricity may be generated by biomass either through the direct combustion of the biomass solids in a steam-Rankine cycle boiler to create steam which then causes a turbine to rotate or, alternatively, in the form of biomass gas (typically about 50 per cent methane and 50 per cent carbon dioxide)[60] which is harvested either through the natural breakdown of the organic material or through chemical processing. Liquefaction and the anaerobic conversion of biomass to either gas or liquid fuels may also be used to capture

54 Tester et al., above n 6, 569; IEA, *Technology Roadmap: Solar Thermal Electricity*, above n 53, 13.

55 Needham, above n 35, 6.

56 Tester et al., above n 6, 562; IEA, *Technology Roadmap: Solar Thermal Electricity*, above n 53, 7.

57 IEA-ETSAP and IRENA, *Concentrating Solar Power*, Technology Brief, IEA-ETSAP Technology Brief E10 (IEA-ETSAP and IRENA, 2013), 21–2.

58 OECD, *OECD Glossary of Statistical Terms* (OECD Publishing, 2008) 501.

59 Donald L Klass, 'Biomass for Renewable Energy and Fuels' in Cutler J Cleveland (ed.), *Encyclopaedia of Energy* (Elsevier Science, 2004) vol 1, 193.

60 Tester et al., above n 6, 437.

and generate electricity.[61] The largest producers of modern bioelectricity generation are China, Brazil, Germany, Japan, the United Kingdom and India.[62] Modern biomass accounted for 5 per cent of total final energy consumed globally in 2016.[63]

Biomass is included in the vast majority of legislative definitions, with 101 countries, as well as the EU and the IRENA, including it in their definitions. Biomass is excluded from the legislative definitions of Palau and St Vincent and the Grenadines (presumably on the basis that that they do not have the processing facilities), as well as Iceland, Palestine, Uzbekistan and Russia. There are a further five countries where it is not conclusive whether or not biomass is incorporated within the legislative definition.

Biomass presents several advantages and disadvantages. One advantage is that primary biomass can be used for energy applications – electricity, heat and transportation fuel. Biomass also has a relatively uniform global distribution relative to fossil fuels and when co-fired with fossil fuels it increases the efficiency of the coal-fired generation plants and reduces the emissions of inorganic matter, sulphur oxides and fossil fuel derived carbon dioxide. In addition, similar to fossil fuels and unlike many other sources of renewable energy, biomass as a feedstock can be stored. However, relative to fossil fuels, biomass has a lower energy density,[64] which makes it expensive to transport when calculated on a joule to cubic metre basis. The other disadvantage that biomass presents is that, unlike other sources of renewable energy, its feedstocks are not freely available. Producing the biomass 'requires a long chain of activities such as planting, growing, harvesting, pre-treatment (storage and drying), upgrading to a fuel, and finally mechanical, thermochemical or biological conversion to an energy carrier'.[65] This means that in addition to the upfront fixed costs for the technology, the project developer must also pay ongoing costs for the biomass feedstocks. An additional concern with biomass is, where the feedstock is not constituted by waste but is rather grown especially for energy production, the potential impact of that on water resources, soil nutrient depletion, biodiversity and the limited availability of land. This raises questions about the impact of using land to grow biomass relative to

61 Klass, above n 59, 204.
62 REN21 Secretariat, above n 30, 71.
63 Ibid 69.
64 Energy content per unit volume or unit mass.
65 Commission of the European Communities, *Communication from the European Commission: Biomass Action Plan* (COM (2005) 628 final), 7 December 2005.

competing land uses, and the resulting impacts on food security, carbon sinks and forestry.[66] For these reasons, the EU has recently proposed the introduction of much stricter sustainability criteria for bioenergy.[67]

If renewable energy must be derived from renewable fuel sources that are regenerated at an equal or faster rate than they are harvested for energy, some (but not all) biomass will qualify as renewable. Where biomass is renewable, it is approximately carbon neutral, with the carbon dioxide extracted from the atmosphere during its growth being released back into the atmosphere during conversion to energy.[68]

2.3.1 *Woody Biomass*

Five countries: Bangladesh, Belarus, Finland, Kenya and Thailand, explicitly refer to the inclusion of wood within the definition of 'renewable energy', while a further four countries: Australia, Nicaragua, the United States and Tajikistan, include wood waste. A number of other countries include woody biomass within their subsidiary definition of 'biomass'.

The inclusion of woody biomass in the definition of renewable energy has long been controversial.[69] This is because of the risk that old growth or non-plantation native forest, which cannot be regenerated at an equal or faster rate than they are harvested for energy, might be captured within the definition.[70] This issue has been raised as a concern in many places that have comparatively

66 IEA, *Technology Roadmap: Delivering Sustainable Bioenergy* (OECD/IEA, 2017) 49; Nicolae Scarlat et al., 'Renewable Energy Policy Framework and Bioenergy Contribution in the European Union – An Overview from National Renewable Energy Action Plans and Progress Reports' (2015) 51 *Renewable and Sustainable Energy Reviews* 969, 971–3; see also General Secretariat of the Council, Proposal for a Directive of the European Parliament and of the Council on the promotion of the use of energy from renewable sources – Analysis of the final compromise text with a view to agreement (21 June 2018) (62bis NEW) <www.consilium.europa.eu/register/en/content/out?&typ=ENTRY&i=LD&DOC_ID=ST-10308-2018-INIT>.

67 Council of the European Union, Commission Staff Working Document: Impact Assessment: Sustainability of Bioenergy accompanying the document proposal for a Directive of the European Parliament and of the Council on the promotion of the use of energy from renewable sources (recast) (COM(2016) 767 final) (SWD(2016) 419 final), 30 November 2016.

68 See e.g. European Commission, 'Achieving Global Leadership in Renewable Energies' (Press Release, 30 November 2016) <http://europa.eu/rapid/press-release_MEMO-16-3987_en.htm>; Daniel M Kammen, 'Renewable Energy, Taxonomic Overview' in Cutler J Cleveland (ed.), *Encyclopedia of Energy* (Elsevier Science, 2004) vol 5, 388.

69 Keith Openshaw, 'Supply of Woody Biomass, Especially in the Tropics: Is Demand Outstripping Sustainable Supply?' (2011) 13 *International Forestry Review* 487.

70 Helena Chum et al., 'Bioenergy' in Ottmar Edenhofer et al. (eds.), *IPCC Special Report on Renewable Energy Sources and Climate Change Mitigation* (IPCC, 2011), 239; Shijie Liu et al., 'A Sustainable Woody Biomass Biorefinery' (2012) 30 *Biotechnology Advances* 785.

high levels of modern biomass use, including the United States,[71] United Kingdom,[72] some of the Nordic and Baltic countries[73] and Australia.[74]

The debate in Australia over the inclusion of old growth forests, and the sustainability of woody biomass as a source of renewable energy more generally, provides a representative example of the concerns raised internationally. This issue has been considered in Australia at a national level by a Senate Committee[75] in its consideration of the Renewable Energy (Electricity) Bill 2000 and Renewable Energy (Electricity) (Charge) Bill 2000,[76] as well as in the subsequent review of that legislation.[77]

In the initial drafting of the Renewable Energy (Electricity) Bill 2000, the definition of renewable energy to be included in the supporting regulations permitted 'biomass by-products of sustainably managed forestry operations'.[78] This was the source of much criticism in the Committee hearings from groups such as Greenpeace and the Australian Conservation Foundation. These groups raised concerns about the negative impact that the logging of native forests would have on biodiversity and noted that any regulatory support given to eligible native forest wastes 'could dramatically alter the economics of using

71 Kelsi Bracmort, 'Biomass: Comparison of Definitions in Legislation Through the 112th Congress' (Report, Congressional Research Service, 14 November 2012); Christine Elizabeth Zeller-Powell, 'Defining Biomass as a Source of Renewable Energy: The Life Cycle Carbon Emissions of Biomass Energy and a Survey and Analysis of Biomass Definitions in States' Renewable Portfolio Standards, Federal Law, and Proposed Legislation' (2011) 26 *Journal of Environmental Law and Litigation* 367; Thomas Walker (ed.), *Massachusetts Biomass Sustainability and Carbon Policy Study: Report to the Commonwealth of Massachusetts Department of Energy Resources* (Manomet Center for Conservation Sciences, 2010).

72 United Kingdom Government, *Biomass as a Renewable Energy Source*, Royal Commission on Environmental Pollution (2004) 28.

73 Inge Stupak, Nicholas Clarke, Anders Lunnan et al., 'Sustainable Utilisation of Forest Biomass for Energy – Possibilities and Problems: Policy, Legislation, Certification, and Recommendations and Guidelines in the Nordic, Baltic, and Other European Countries' (2007) 31 *Biomass and Bioenergy* 666.

74 Senate Environment, Communications, Information Technology and the Arts References Committee, Commonwealth of Australia, *Report of the Senate Environment, Communications, Information Technology and the Arts References Committee into the Renewable Energy (Electricity) Bill 2000 and Renewable Energy (Electricity) (Charge) Bill 2000* (2000).

75 Ibid.

76 Ibid.

77 See e.g. Dick Warburton et al., *Renewable Energy Target Scheme: Report of the Expert Panel* (Commonwealth of Australia, 2014) 84–7 <http://webarchive.nla.gov.au/gov/20150401162302/https://retreview.dpmc.gov.au/>.

78 Senate Environment, Communications, Information Technology and the Arts References Committee, above n 74, 12.

native forest energy production'.[79] The Committee also took into account that this approach had not been adopted under the New South Wales State Government's Green Power Scheme. The Sustainable Energy Development Authority (SEDA), which administered the NSW Green Power Scheme, stated in its guidelines that

> utilisation of waste derived from sustainably harvested plantation forest is generally accepted under Green Power. These wastes should not be sourced from plantations that clear, or have cleared after 1990, existing old-growth native forests.[80]

Indeed, even the utilisation of waste products from regrowth in native forests for Green Power under this Scheme was deemed to be a highly sensitive issue. The Australian Greenhouse Office, the relevant regulator at the time of the drafting of the Bill, admitted that while native forests, including some old-growth forests, would be subject to logging for woody biomass under the relevant regulations, 'all relevant approvals – Commonwealth, State or local – must be obtained for accreditation under this measure'.[81] However, the Committee rejected the view that this would be an adequate protection to ensure that biodiversity conservation values were preserved and that logging was undertaken sustainably. In particular, the Committee noted that the Australian Greenhouse Office did not have independent expertise in forestry policy and nor did it have resources to devote to such matters. This led to the first recommendation of the Committee Report being 'that non-plantation native forest wood products and wood waste be specifically excluded from the list of eligible renewable energy sources'.[82]

This recommendation was adopted in regulation 8 of the Renewable Energy (Electricity) Regulations 2001, which placed a limitation of the definition of woody biomass incorporated within the definition of 'renewable energy'. This meant that only wood from plantation forests that could be harvested and then regenerated without significant loss of biodiversity was deemed to be an eligible renewable energy source. It also meant that the economics of the logging of old-growth or non-plantation native forests were not distorted by the RET scheme.

79 Ibid 12–13.

80 Sustainable Energy Development Authority, *Green Power Briefing: Greenpower and Wood Wastes* (unknown) 1 quoted in Senate Environment, Communications, Information Technology and the Arts References Committee, above n 74.

81 Senate Environment, Communications, Information Technology and the Arts References Committee, above n 74, 108.

82 Ibid vii.

This issue was revisited in 2014 in the Review of the Renewable Energy Target (RET) Scheme following the new conservative Federal Government's election promise to reinstate native forest wood waste as an eligible renewable energy source under the RET. Not surprisingly, forestry bodies and three State Governments with significant forestry industries supported this change,[83] while environmental and community groups raised concerns about the potential impacts of such a change.[84] In response, the Expert Panel stated that

> concerns about the sustainable logging of native forests [were] . . . outside of the scope of this review, [and that] in contrast to disposing of native forest wood waste by either incineration or allowing waste to decompose, utilising wood waste in a power station may be a more efficient use of resources and lead to lower CO_2-e emissions by reducing the use of gas or coal.[85]

As a result, the Expert Panel recommended that native forest wood waste (from any source) should be supported under the RET Scheme.[86] This change has been reflected in amendments to s 8 of the Renewable Energy (Electricity) Regulations 2001, though a significant limitation remains, the production of biomass cannot be the primary purpose for which the native forest was harvested.

As stated above, it is broadly accepted that biomass is a renewable source of electricity generation. However, the use of non-plantation native forest wood products and wood wastes as a source of biomass would arguably cause biodiversity losses which could not be regenerated within a plausible timescale relative to the use of the harvested wood. As a result, the inclusion of old growth native forests within the definition of renewable energy should be robustly questioned.

2.3.2 *Traditional Biomass*

Both modern and traditional biomass are significant contributors to the global energy supply, with traditional forms of biomass providing 7.8 per cent of the total final energy consumed globally in 2016.[87] 'Traditional biomass' is predominantly used in developing countries for cooking and heating where charcoal, animal dung and firewood have long been traditional fuel sources. Indeed, Chum et al. have found that 60 per cent (IEA accounted) to

83 New South Wales, Victoria and Tasmania.
84 Warburton et al., above n 77, 86.
85 Ibid 87.
86 Ibid.
87 REN21 Secretariat, above n 30, 69.

70 per cent (including the unaccounted informal sector) of total biomass use occurred in rural areas of developing countries.[88]

Some low to lower-middle income developing countries such as Kenya specifically include forms of traditional biomass such as charcoal within their legislative definition of 'renewable energy'.[89] This reflects the fact that charcoal is a major energy source for the Kenyan population, providing 82 per cent of household energy in urban areas, employing almost 900,000 people and contributing an estimated $US1.6 billion to the Kenyan economy annually.[90] At the same time, the use of traditional biomass in Kenya has led to widespread environmental damage, including 'leaving less than 2% of Kenya in forest cover'[91] due to the widespread failure to replant the harvested forest. In 2018, the environmental degradation associated with charcoal use prompted the Kenyan Government to again issue a short-term ban on its production in some counties, as well as placing restrictions on its transportation and trading.[92] This meant that contradictory messages were being sent with charcoal simultaneously being listed as an eligible source of renewable energy within the Energy Act, while also being banned in a number of counties. As a result of these bans, there has been an increased push for the development of a sustainable charcoal industry. A few of the ways in which the charcoal industry can be made more sustainable include:

- replanting trees after an area has been harvested;
- growing the most appropriate species of trees for charcoal production;
- using more efficient kilns in the production stage.[93]

Other efforts to try and improve the sustainability of traditional biomass include carbonising agricultural and paper waste rather than wood fuel to make char, which can then be compressed into blocks to form a more efficient

88 Chum et al., above n 70, 246.

89 The Energy Act 2006, Art 2 (Kenya).

90 Hannah Wanjiru et al. *How Kenya can Transform the Charcoal Sector and Create New Opportunities for Low-carbon Rural Development*, Stockholm Environment Institute and UNDP Discussion Brief (Stockholm Environment Institute – Africa World Agroforestry Centre, 2016) 1.

91 BrightGreen Renewable Energy, Kenya, About Us (2016) <www.brightgreenkenya.co.ke/aboutus.html>.

92 Mary Njenga, *Banning Charcoal Isn't the Way to Go. Kenya Should Make It Sustainable*, The Conversation (16 May 2018) <https://the conversation.com/banning-charcoal-sint-the-way-to-go-kenya-should-make-it-sustainable-9560>.

93 See e.g. Wanjiru et al., above n 90; Njenga, above n 92; Hudson Gumbihi, 'Nairobi City Residents Feeling the Heat of the Charcoal Ban', *The Standard* (Online) (6 April 2018) <www.standardmedia.co.ke/business/article/2001275856/lifestyle-change-as-coat-of-charcoal-doubles-in-nairobi>.

and less polluting form of charcoal.[94] The development of these improved processes and technologies will be significant for sub-Saharan Africa, which produces 62 per cent of the global charcoal supply.[95]

Despite these developments, it remains rare for traditional biomass to be included within definitions of renewable energy. This is due to the negative health impacts that result from burning traditional biomass in poorly ventilated spaces and its inefficiency as a fuel source. Neither the IRENA nor the EU recognise traditional biomass, nor does any developed country. Indeed, it is far more common for countries explicitly to exclude traditional biomass from their definitions than it is to include it. China takes this approach in their Renewable Energy Law (2005), which states:

> This Law shall not apply to the utilization of straws or stalks, firewood or dung in the form of direct burning through an inefficient cooking range.[96]

2.4 LANDFILL GAS, SEWAGE TREATMENT GAS AND BIOGAS

Landfill gas, sewage treatment gas and biogas are also well supported within the legislative definitions of 'renewable energy', being included in the legislative definitions of 101 countries, as well as in the EU and the IRENA definitions. Similar to biomass, landfill gas, sewage treatment gas and biogas are excluded from the legislative definitions of Iceland, Palau, Palestine, Russia, St Vincent and the Grenadines and Uzbekistan. There are a further five countries where it is not conclusive whether landfill gas, sewage treatment gas and biogas are incorporated into the legislative definition or not.

Landfill gas, sewage treatment gas and biogas are all formed as a by-product of the process by which anaerobic bacteria break down rotting organic matter in the absence of oxygen using their natural digestive and fermentation processes.[97] This process (sometimes also called 'wet biomass conversion') converts the complex organic material to simple organic material,[98] and also produces gas, which is made up of approximately 50 per cent methane and 50 per cent carbon dioxide, as well as other trace gases.[99] When mixed with oxygen, the methane gas captured in this process can be burned for fuel

[94] BrightGreen Renewable Energy, above n 91.

[95] Njenga, above n 92.

[96] Ch 1, Art 2.

[97] Andrew J Waskey, 'Landfill Methane' in Dustin Mulvaney (ed.), *Green Power* (SAGE, 2011) 237.

[98] Ibid.

[99] Tester et al., above n 6, 437.

(thermal energy) or power a turbine (mechanical or electrical energy).[100] It can also be used in cogeneration.[101] In sufficient quantities, this gas may also be condensed into compressed natural gas or even liquefied natural gas (LNG), which may then be used as an alternative vehicle fuel.[102]

The anaerobic digestion plants or bio-digesters can be used to facilitate the anaerobic digestion of a wide range of biodegradable materials, including municipal waste (landfill gas), sewage (sewage treatment gas) and other organic matter such as vegetation, manure, abattoir waste, green waste and compost (biogas).[103] Landfill gas, sewage treatment gas and biogas have two key advantages. First, they aid in the disposal of organic matter in a cleaner and more efficient manner, producing three useful by-products:

1. methane: which can be used for electrical generation or burned for fuel;
2. carbon dioxide: which can be used for industrial processes such as oil well enhancement; and
3. mineral and nitrogen-rich fertiliser: which can then be used to replace essential nutrients into the soil.[104]

Second, providing there is a continuous source of biodegradable feedstock, these gases provide a cheap and reliable source of renewable energy. In the case of landfill gas and sewage treatment gas, the economics are particularly favourable, as the cost of the feedstocks is actually negative because the waste collectors are paid for waste removal.[105]

The production of biogas is also often beneficial to the health of local communities, as they remove the need to use the traditional disposal processes in developing countries of burning the waste or animal dung which cause air pollution and release particulate matter that are detrimental to human health.[106] It also provides a means of dealing with some of the most prolific sources of anthropogenic emissions in the forms of methane gas and carbon dioxide by preventing their release into the atmosphere and diverting them to other uses.[107]

100 Andrew J Waskey, 'Biogas Digester' in Dustin Mulvaney (ed.), *Green Power* (SAGE, 2011) 36.

101 OECD, *Medium-Term Renewable Energy Market Report* (OECD/IEA, 2012) 138; IEA, *Renewable Energy: RD&D Priorities*, above n 37, 45.

102 Stephen Larkin, Janet Ramage and Joanathan Scurlock, 'Bioenergy' in Godfrey Boyle (ed.), *Renewable Energy: Power for a Sustainable Future* (2nd edn, Oxford University Press, 2004) 128.

103 Waskey, 'Biogas Digester', above n 100, 35–6.

104 Tester et al., above n 6, 438; Waskey, 'Biogas Digester', above n 100, 36.

105 Tester et al., above n 6, 443.

106 Waskey, 'Biogas Digester', above n 100, 37.

107 Ajayi Oluseyi Olanrewaju, 'Biomass Energy' in Dustin Mulvaney (ed.), *Green Power* (SAGE, 2011) 40.

2.5 HYDROPOWER

Hydropower is the largest source of renewable energy generation globally, accounting for approximately 70.3 per cent of renewable electricity production, 16 per cent of electricity generated from all sources and 2.5 per cent of the world's total energy production.[108] In fact, following the closure of some nuclear power plants after the Fukushima disaster, all of the top ten largest electricity generation stations in the world are powered by hydropower.[109] As a result of its long established use, hydropower is considered by the IEA as a 'mature technology'.[110]

Hydropower generates electricity by passing water (either run of the river or from stored capacity in dams) from a higher elevation to a lower elevation through hydraulic turbines, under the influence of the earth's gravitational field.[111] These turbines drive a rotating shaft that in turn powers an electric generator.[112]

2.5.1 *Small-Scale Hydropower*

Hydropower systems can range from pico-size (50W to 5kW) to large-scale operations that generate more than 100MW to be fed into the grid.[113] The classification of the size of hydropower plants varies considerably by country, as there is not an internationally recognised standard. For example, the maximum generating capacity for a hydropower system to be considered small-scale within the national renewable energy laws under study ranges from 3MW in Panama to 30MW in Malaysia and Tajikistan. The following table highlights the classification system adopted by Renewable Energy and Energy Efficiency Partnership/United Nations Industrial Development Organization (UNIDO):

108 IEA, *Renewables Information (2017 Edition)* (OECD/IEA, 2017) vii, ix.

109 US Energy Information Administration, *The World's Nine Largest Operating Power Plants Are Hydroelectric Facilities* (18 October 2016) <www.eia.gov/todayinenergy/detail.php?id=28392&src=email>.

110 IEA, *Hydropower* (2018) <www.iea.org/topics/renewables/hydropower/>.

111 Tester et al., above n 6, 520; see also IEA, *Technology Roadmap: Hydropower* (OECD/IEA 2012) 11–14.

112 Tester et al., above n 6, 526.

113 Renewable Energy and Energy Efficiency Partnership and United Nations Industrial Development Organization, 'Module 7: Renewable Energy Technologies' in *REEEP/UNIDO Training Package on Sustainable Energy Regulation and Policymaking for Africa* (REEEP/UNIDO, 2008) 7.21 <http://africa-toolkit.reeep.org/>.

TABLE 2.1 *Classification of Hydropower Projects by Size*[114]

Type of hydropower project	Details
Pico-hydro	50W to 5kW – usually for remote communities and individual households. Applications include battery charging or food processing.
Micro-hydro	5kW to 100kW – usually provide power for a small community or rural industry in remote areas away from the grid
Mini-hydro	100kW to 1MW – either stand-alone schemes or more often feeding into a grid
Small-hydro	1 MW to 10MW or 20MW – definitions vary; Europe tends to use 10MW as a maximum, China uses 20MW and Brazil 30MW. Usually feeding into a grid.
Medium-hydro	10 or 20MW to 100MW – usually feeding into a grid
Large-hydro	More than 100MW and usually feeding into a large electricity grid[115]

The classification of the size of the hydropower plant is important. Over 25 per cent of countries with renewable energy laws do not support the inclusion of large-scale hydropower as a renewable resource, often due to its environmental and social impacts. In contrast, small-scale hydropower, which tends to avoid the construction of dams, rather relying on the natural flow of rivers, enjoys high levels of regulatory support. 102 countries included small-scale hydropower within their legislative definitions of renewable energy, with the EU and the IRENA also including it within their legislative definitions. Finland does not explicitly support hydropower within their primary renewable energy law, but may be inferred to have incorporated it by virtue of having to transpose the relevant EU Directives on renewable energy. Palestine, Paraguay and Tunisia exclude the inclusion of all forms of hydropower. It is not clear whether small-scale hydropower is included or excluded from the legislative definitions of renewable energy in a further seven countries.

Small-scale hydropower offers a number of advantages as a form of renewable energy. Other than the initial installation costs, the running

114 Ibid.

115 Note however that the IRENA defines any facility that has a generating capacity greater than 10MW as a large-scale hydropower facility.

and maintenance costs are generally low when compared to other energy sources.[116] In addition, the anticipated lifespan of small-scale hydropower systems is over thirty years,[117] with electricity being provided at a fairly constant rate (subject to seasonal and climatic variability) throughout that period.[118] That said, small-scale hydropower has certain disadvantages, primarily due to the fact that its development is subject to finding suitable geographic locations close to where the electricity will be used.[119] Further, the ability to expand the operation of a small-scale hydropower plant is often limited post-construction due to the natural constraints of being located in a small stream or river.[120]

A further advantage of small-scale hydropower is that, relative to large-scale hydropower and many fossil fuel sources of electricity generation, it has fewer adverse environmental impacts. However, small-scale hydropower plants may still cause problems 'when water levels in reservoirs change abruptly to meet electricity demands or in times of low flow, the short stretch of by-passed river can run dry, which might dry out aquatic organisms'.[121] This may lead to environmental impacts such as riverbed erosion and loss of biodiversity, as well as water usage impacts for communities downstream.

2.5.2 *Large-Scale Hydropower*

As stated above, the inclusion of large-scale hydropower within the definition of renewable energy has lower levels of national acceptance than small-scale hydropower. Seventy-seven countries included large-scale hydropower within their legislative definitions of renewable energy, with the EU and the IRENA also including it within their legislative definitions. Twenty-seven countries including Austria, Greece, Panama, Senegal, Tajikistan and Thailand specifically exclude large-scale hydropower. It is not clear whether large-scale hydropower is included or excluded from the legislative definitions of renewable energy in a further nine countries.

116 Maria Jose Descalzo et al., 'Small-Scale Hydropower in Europe: Legal Issues' in Matt Bonass and Michael Rudd (eds.), *Renewables: A Practical Handbook* (Globe Business Planning, 2010) 210–11.

117 REEEP/UNIDO, above n 113, 7.22.

118 Ibid 7.24.

119 Descalzo et al., above n 116, 210–11; Tester et al., above n 6, 533.

120 REEEP/UNIDO, above n 113, 7.23.

121 Ibid.

The inclusion of large-scale hydropower within the definition of 'renewable energy' is controversial for four reasons:

1. the environmental impact associated with the construction of a large dam or series of dams required for large-scale hydropower projects;
2. the social impact of displacing local populations when areas are flooded during the construction process;
3. there are natural constraints on the location of new dams; and
4. to include hydropower, which has already benefited over a long period of time from significant government subsidies in the definition of 'renewable energy', may provide the existing hydropower projects with super profits unless they are explicitly excluded. Further, in most countries the provision of subsidies is likely to lead to very limited additional generation capacity being installed due to the natural constraints on the location of new large dams.

2.5.2.1 The Environmental Impact of Hydropower

Due to the requirement of large hydropower projects such as the Three Gorges Dam in China, Itaipu Dam in Brazil and the Snowy Hydro Electric Scheme in Australia, of either a large dam or a series of dams, the construction of large-scale hydropower projects has a significant impact on the environment. A number of environmental impacts from large-scale hydropower projects have been cited as problematic, including:

1. the displacement and resultant loss of biodiversity when the habitat of flora and fauna is flooded to build the dams required for these large-scale projects;[122]
2. 'the decay of plant life in the dam reservoirs produces significant amounts of carbon dioxide';[123]
3. dams tend to collect sediment, which not only may reduce their effectiveness as a source of hydropower over time[124] but may also 'reduce the amount of silt being carried downstream ... reducing the fertility of downstream soil';[125]

[122] Louisa Fitz-Gerald and Paul Curnow, 'Australian Case Study' in Leslie Parker et al. (eds.), *From Debate to Design: Issues in Clean Energy and Climate Change Law and Policy: A report on the Work of the REIL Network 2007–2008* (Yale School of Forestry & Environmental Studies, 2008) 115; Kammen, above n 68, 404.

[123] Ottinger and Bradbrook, above n 1, 6.

[124] Ibid.

[125] Fitz-Gerald and Curnow, above n 122, 136–7.

4. the requirement of large stores of water required for large-scale hydropower projects at a time when national water shortages are becoming more widespread;[126]
5. the dams needed for the generation of hydropower impact on local ground water flows, altering the ecology of surrounding areas;[127]
6. damming can also have a negative impact on water quality,[128] reduce fish populations[129] and 'contribute to the spread of water-related diseases';[130] and
7. areas that have been dammed are more susceptible to natural disasters such as flooding, landslides and rock falls, potentially endangering local communities.[131]

This has led to the argument that the environmental impact of hydropower may be sufficiently significant that from a policy perspective governments ought to consider whether they are a sustainable development, as well as being 'renewable'.[132] Indeed, Kammen has argued that:

> properly addressing these issues would result in an enormous escalation of the overall costs of producing hydropower, making it far less competitive than is usually stated.[133]

2.5.2.2 The Social Impact of Hydropower

The second reason why the inclusion of large-scale hydropower has been controversial is that the process of flooding often results in the displacement of significant numbers of the local population. For example, during the construction of the Three Gorges Dam Project in China between 1994 and 2006, some 13 cities, 140 towns and 1,350 villages were

126 New South Wales Government Office of Water, *Final Report: Revision of Snowy Hydro Water Licence: Requirements for the release from Snowy Hydro's storages of water 'deficit' accumulated under dry inflow sequence during drought years: Response to submissions June 2011* (NSW Office of Water, 2011) 8; REN21 Secretariat, above n 30, 43.

127 Fitz-Gerald and Curnow, above n 122, 136–7.

128 Kammen, above n 68, 404.

129 Ibid 404.

130 Ibid 404.

131 ABC News, 'Dam Collapse Highlights Risks to Communities as Laos Seeks to Become HydroElectricity Hub', *ABC News* (online), 3 August 2018 <www.abc.net.au/news/2018-08-03/dam-collapse-highlights-risks-to-communities-in-laos/10071836>; BBC China, 'China's Three Gorges Dam May Displace Another 100,000', *BBC World News* (online), 18 April 2012 <www.bbc.co.uk/news/world-asia-china-17754256>.

132 Andrii Gritsevskyi, *Renewable vs. Non-renewable Energy Sources, Forms and Technologies* (UN Statistical Commission, unknown) 5–6.

133 Kammen, above n 68, 404.

submerged,[134] displacing an estimated 1.3 million people.[135] This displacement has dislocated the local population from their traditional homes and land, often removing them from subsistence agriculture and separating local communities. This is a problem that is often inadequately addressed.

2.5.2.3 Natural Constraints on the Growth of Large-Scale Hydropower

The third reason why large-scale hydropower projects are controversial is that due to their scale, there are often natural constraints on the construction of new dams. That is, there needs to be natural water flows from higher points of elevation to lower levels of elevation, as well as land that can be viably flooded and contained in impoundment dams. This means that large-scale hydropower projects are not 'renewable' unless there are the right environmental conditions.

2.5.2.4 The Super Profits and Additionality Problems

Hydropower is already one of the cheapest forms of energy available due to its long infrastructure life compared with other energy projects, and low maintenance costs.[136] Further, due to its historical development largely occurring in the immediate pre- and post-World War II period, the hydropower sector has already benefited from extensive public subsidies.[137] To incorporate hydropower into the definition of 'renewable energy' adopted by the primary legislative instruments in each of the jurisdictions studied would perversely provide the owners of hydropower projects with super profits. This problem was recognised in debates within the US Senate Energy Committee by Mr Leon Lowery when he stated:

> If you try to assign a sort of conceptual definition, you find yourself in strange places [...] Anyone would acknowledge that hydropower is renewable, but do we want to give credits to the Grand Coulee Dam? To do so, would give hydropower – which already benefits from rich federal subsidies that make it some of the cheapest energy available – the same status as solar or wind technologies.[138]

Also, due to the natural constraints described above, it is unlikely that the provision of any regulatory support would lead to the development of many new large-scale projects throughout the world. Thus inclusion of large-scale

[134] BBC China, above n 131.

[135] Michael Bristow, 'China Acknowledges Three Gorges Dam "Problems"', *BBC World Service* (online), 19 May 2011 <www.bbc.co.uk/news/world-asia-pacific-13451528>.

[136] Ottinger and Bradbrook, above n 1, 6.

[137] Tester et al., above n 6, 535.

[138] Felicity Barringer, 'With Billions at Stake, Trying to Expand the Meaning of "Renewable Energy"', *New York Times* (New York), 25 May 2009.

hydropower is not likely to lead to many new additional projects being developed but rather (unless it is appropriately targeted) is more likely to lead to super profits being generated for existing projects.

2.5.2.5 Should Large-Scale Hydropower Be Defined as 'Renewable Energy'?

Despite the varied negative impacts of large-scale hydropower outlined above, and recognising that all forms of energy produce externalities,[139] it does have a number of positive externalities that may support its inclusion within the definition of 'renewable energy'. Due to the maturity of the technology, the lack of polluting emissions produced, the little maintenance required and the cost comparisons with other forms of energy using current accounting methods, it can be argued that large-scale hydropower produces relatively plentiful, cheap and reliable energy, particularly when compared with other energy forms. It is for this reason that while twenty-seven countries exclude large-scale hydropower projects from their national renewable energy laws, the vast majority of countries with such laws have chosen to include hydropower within their definition of 'renewable energy'. That said, where regulatory support is provided to large-scale hydropower, it is often quite limited for the reasons outlined above when compared with other energy technologies that need more support to encourage their research, development and subsequent commercialisation.

2.5.3 *Pumped Hydropower*

A recent development that is emerging with the increased deployment of energy storage technologies, particularly pumped hydropower, is that some countries such as Albania and Taiwan are beginning specifically to exclude it from their definition of 'renewable energy'. This approach has also been adopted by the EU in both the current, and the proposed re-cast, Renewable Energy Directive, which specifically states:

> ... electricity produced in pumped storage units from water that has previously been pumped uphill should not be considered to be electricity produced from renewable energy sources.[140]

[139] Peter Bickel and Rainer Friedrich, *ExternE Externalities of Energy* (DG Research European Commission, 2005).

[140] Directive 2009/28/EC of the European Parliament and of the Council of 23 April 2009 on the promotion of the use of energy from renewable sources and amending and subsequently repealing Directives 2001/77/EC and 2003/30/EC [1993] OJ L 140/16, recital 30; General Secretariat of the Council of the European Union Proposal for a Directive of the European Parliament and of the Council on the promotion of the use of energy from renewable sources – Analysis of the final compromise text with a view to agreement, above n 66, 14.

This is due to the hydraulic and electrical losses incurred in pumping water from a lower reservoir to an upper reservoir, meaning that pumped hydropower facilities are net consumers, as opposed to producers, of electricity. Despite this, pumped hydropower plays an important role in providing storage, ancillary services and system balancing, and is often highly efficient.[141]

2.6 GEOTHERMAL ENERGY

Geothermal energy involves utilising the thermal energy, initially generated by friction from the continental plates rubbing against one another and the decay of small amounts of naturally occurring radioactive elements,[142] stored in the Earth's crust.[143] There are five forms of geothermal energy: volcanic or magmatic reserves; vapour dominated systems; geopressured systems; geothermal hot springs/aquifers;[144] and hot fractured or dry rocks.[145] The last two forms of geothermal energy 'resources are widespread throughout the world'[146] and thus will be the focus of this discussion.

Geothermal energy has been lauded for its low greenhouse gas emissions when compared to fossil fuel generation sources. Geothermal power plants produce very minimal levels of nitrous oxide and sulphur gases, and only a sixth of the carbon dioxide that an efficient natural gas generation facility produces.[147] Despite this, there has been much debate about whether geothermal energy is a renewable energy source.[148]

141 Energy Storage Association, *Pumped Hydroelectric Storage* (2018) <http://energystorage.org/energy-storage/technologies/pumped-hydroelectric-storage>.

142 Kammen, above n 68, 405.

143 Paul Breeze, 'Geothermal Power' in Paul Breeze (ed.), *Power Generation Technologies* (Elsevier Science, 2014) 243.

144 Also sometimes referred to as 'hydrothermal energy'. For more discussion of this point, please see the discussion of hydrothermal energy under the ocean, riverine and tidal energy section.

145 Adrian Bradbrook and Anita Rønne, 'New Advances in Geothermal Energy Law: A Comparative Analysis' in Donald Zillman et al. (eds.), *The Law of Energy Underground: Understanding New Developments in Subsurface Production, Transmission, and Storage* (Oxford University Press, 2014) 309, 311.

146 Adrian Bradbrook, 'The Role of the Common Law in Promoting Sustainable Energy Development in the Property Sector' in Aileen McHarg et al. (eds.), *Property and the Law in Energy and Natural Resources* (Oxford University Press, 2014) 391, 404.

147 Office of Energy Efficiency & Renewable Energy, United States of America (2018) 'Geothermal FAQs' <www.energy.gov/eere/geothermla/geothermal-faqs>.

148 See e.g. Valgardur Stefansson, 'The Renewability of Geothermal Energy' (Proceedings of the World Geothermal Congress, Kyushu-Tohoku, 2000); Silja Ran Siguroardottir et al., 'Optimising Revenue of a Geothermal System with Respect to Operation and Expansion' (Proceedings of the World Geothermal Congress, Bali, 2010); IEA, *Technology Roadmap: Geothermal Heat and Power* (OECD/IEA, 2011) 6.

Ninety-six countries included geothermal energy within their legislative definitions of renewable energy, with the EU and the IRENA also including it within their legislative definitions. The Netherlands does not provide support to hot dry rocks but rather limits its support to geothermal aquifers. Thirteen countries exclude support for geothermal energy from their legislative definition of renewable energy. It is not clear whether geothermal is included or excluded from the legislative definitions of renewable energy in a further four countries.

One of the reasons for the strong support is that there is enormous geothermal potential globally, with Kammen stating that:

> Scientists estimate that just 1% of the heat contained in the uppermost 10 km of Earth's crust is equivalent to 500 times the energy contained in all of Earth's oil and gas resources.[149]

To take Australia as an example, 'preliminary work by Geoscience Australia suggests a potential [hot fractured rock] resource equivalent to 26,000 years of Australia's energy use at 2005 levels'.[150] From a more cynical perspective, the fact that hot dry rock technology is also yet to be successfully commercialised on a large scale means that it can be safely included within the legislative definitions with limited potential costs to countries in the short to medium term.

2.6.1 *Geothermal Hot Springs and Aquifers*

The first method of capturing geothermal energy involves using hot springs or geothermal aquifers. Most geothermal power stations use the high water temperatures of between 100°C and 300°C[151] found in naturally occurring geothermal hot springs or reservoirs to power either flash or dry steam condensing turbines or to power a binary-cycle plant.[152] In 2015, geothermal power provided more than 50 per cent of the electricity generated in Kenya, and more than 25 per cent of the electricity generated in Iceland, the Philippines

149 Kammen, above n 68, 405.

150 Barry A Goldstein et al., 'The National Outlook – Australia's Hot Rocks' (Paper presented at the Proceedings of the Sir Mark Oliphant International Frontiers of Science and Technology Australian Geothermal Energy Conference, Melbourne, August 2008) 13 <www.ga.gov.au/image_cache/GA11825.pdf> .

151 IEA, *Technology Roadmap: Geothermal Heat and Power*, above n 148, 9.

152 Barry A Goldstein, Gerardo Hiriart et al., 'Geothermal Energy' in Ottmar Edenhofer et al. (eds.), *IPCC Special Report on Renewable Energy Sources and Climate Change Mitigation* (IPCC, 2011) 401.

and El Salvador.[153] By 2016, there was 12.7GW of geothermal power capacity installed[154] in plants in twenty-five countries.[155]

Geothermal aquifers tend to be used when there is insufficient heat to generate steam to power a turbine for electricity energy. Geothermal aquifers provide heat for residential and commercial applications in more than seventy countries.[156]

2.6.2 *Hot Fractured Rock Technology*

Hot fractured rock technology involves drilling boreholes into hot rocks 'to facilitate the injection of water, which passes through fractures in the rock to extraction boreholes and returns to the surface as steam'.[157] Two preconditions are necessary for a successful hot fractured rock project. The first precondition is the ability to drill economically and safely to a sufficient depth into the hot dry rock. This has proved problematic, with hot fractured rock technology still at an experimental stage due to difficulties associated with drilling at such a deep depth into hot dry rock. The second precondition is that there must be fractures that allow the water to pass through to the extraction bores. Thus, while geothermal energy is available in essentially inexhaustible quantities, the requirement for fractured rock at sufficiently high temperatures poses difficulties. For example, geothermal energy is unevenly distributed; it is seldom concentrated (for example, while Australia has naturally occurring sites within five kilometres of the earth's surface which reach temperatures over 250°C, the United Kingdom only has a few sites); and it 'is often found at depths too great to be exploited industrially and economically'.[158] It is also one of the least commercialised renewable energy technologies, with it likely to be at least ten to fifteen years before hot dry rock technologies are commercially viable.[159] Due to its stage of development, the high costs involved in the

153 Bertani, Ruggero 'Geothermal Power Generation in the World 2010–2014 Update Report' (Paper presented at the Proceedings of the World Geothermal Congress 2015, Melbourne, Australia, 19–25 April 2015) 2, 3 <https://pangea.stanford.edu/ERE/db/WGC/papers/WGC/2015/01001.pdf>.

154 IRENA, *Geothermal Power: Technology Brief* (IRENA, 2017) 5.

155 Tim Williamson, '2017 *Geothermal Energy International Forum*' (Paper presented at the 2017 Geothermal Energy International Forum, Washington DC, 7 March 2017) <www.irena.org/eventdocs/2017%20Geothermal%20Energy%20International%20Forum_2.03.2017.pdf>.

156 Sylvia Harrison, 'Geothermal Resources' in Michael B Gerrard (ed.), *The Law of Clean Energy – Efficiency and Renewables* (American Bar Association, Section of Environment, Energy and Resources, 2012) 423, 425.

157 Needham, above n 35, 8.

158 Kammen, above n 68, 405.

159 Breeze, above n 143, 248.

technology, and in the absence of very high petroleum prices, hot fractured rock geothermal energy will need to rely on significant regulatory support if it is to be effectively commercialised and its potential realised.

2.6.3 *The Environmental Impact of Geothermal Energy*

In addition to the challenges detailed above with hot fractured rock technology, geothermal energy also poses three environmental risks which have led some institutions such as the United Nations (UN) to question whether geothermal energy should be defined as 'renewable energy'.[160] The first risk is that some use may harm the amount of energy that is effectively available at some locations, meaning that it could in fact be deemed to be exhaustible within a defined location.[161] While natural variations of geothermal resources occur over hundreds or even thousands of years, 'man-induced processes lead to variations within shorter time scales, typically in the range of decades'. In some cases, over-exploitation and/or improper reinjection permanently damages the geothermal resource within a geographic area. This led Brown et al. to argue that 'it must be tapped slowly enough so as not to deplete the accessible reservoir of heat, and thus be truly renewable'.[162] However, others such as Kozloff and Dower have argued that a perspective of 300 years or more of continuous production is adequate for an energy fuel to be considered as renewable, since technical advances during that time will have rendered today's perspective obsolete.[163] Arguably, if this approach is to be taken, given the prediction by Geoscience Australia that Australia has at least 26,000 years' worth of geothermal potential according to 2005 energy levels,[164] the argument that local geothermal resource may be exhausted in a defined location should not prevent geothermal energy from being deemed to be renewable.

The second risk is that geothermal fluids contain variable concentrations of gases, largely methane, nitrogen, carbon dioxide and hydrogen sulphide, with smaller proportions of ammonia, radon and boron, and trace amounts of mercury and arsenic.[165] This risk is largely minimised through the gas and

160 Gritsevskyi, above n 132, 8.

161 Ibid; see also, Breeze, above n 143, 247.

162 Lester R Brown, Christopher Flavin and Sandra Postel, *Picturing a Sustainable Society: State of the World* (Worldwatch Institute, 1990) quoted in Tester et al., above n 6, 487.

163 Keith L Kozloff and Roger C Dower, *A New Power Base – Renewable Energy Policies for the Nineties and Beyond* (World Resources Institute, 1993) cited in Tester et al., above n 6, 487.

164 Goldstein et al., 'The National Outlook – Australia's Hot Rocks', above n 150.

165 Kammen, above n 68, 407; Harrison, above n 156, 436.

water released by the projects either being reinjected into the subsurface of the earth's crust (normally into steel encased exhausted bore holes) or collected and chemically altered for safe disposal.[166] Particularly when compared with the emissions and pollutants expelled from fossil fuel powered electricity generation stations, the environmental impacts of geothermal energy are relatively low and can be managed with the aid of existing technologies. However, this concern has been explicitly noted by the EU in the Renewable Energy Directive, which states that they 'should only facilitate the deployment of geothermal energy with low environmental impact and resulting in greenhouse gas emission saving compared to conventional sources'.[167]

The third risk is the potential loss of thermophile biodiversity, that is, the microbial organisms that have adapted to cope with the extreme heat and chemical compositions of geothermal resources. While comparatively little is known about these organisms and their environments, they have already proven valuable to scientific and medical research.[168] This means that the potential impacts of geothermal energy generation are poorly understood and thus a precautionary approach may be warranted.

In conclusion, despite the fact that geothermal energy leads to the slow exhaustion of the resource within defined locations and hot dry rock geothermal energy is still largely to be commercialised, geothermal energy possesses significant potential as a form of renewable energy. As such, and in accordance with the widely adopted international position, geothermal energy is commonly considered to be a renewable source of electricity generation.

2.7 OCEAN AND RIVERINE ENERGIES

Ocean and riverine energies, being energy derived from tides, waves, marine currents, hydrothermal layering and osmotic energy,[169] are some of the least commercially developed sources of renewable energy. By the end of 2017, there was only 529MW of installed ocean energy capacity globally, with the vast majority of this being tidal power.[170] This low level of deployment is due

166 Harrison, above n 156, 436–7.

167 General Secretariat of the Council of the European Union, Proposal for a Directive of the European Parliament and of the Council on the promotion of the use of energy from renewable sources – Analysis of the final compromise text with a view to agreement, above n 66, 19, Preamble, recital 31a.

168 Tiffany Grant, 'California Geothermal Law and Its Impacts on Thermophile Biodiversity' (2005) 6(2) *Engage* 52.

169 Needham, above n 68, 10.

170 REN21 Secretariat, above n 30, 88.

to the high capital costs currently associated with these technologies,[171] as well as concerns about their vulnerability in storms[172] and their impacts on navigation[173] and the environment.[174]

The identification and classification of ocean and riverine energies in the legislative definitions of renewable energy is often difficult due to the terminology used by some countries in their legislation. Phrases such as 'natural water flows', 'water energy', 'hydraulic energy' and 'watercourses' are often ambiguous in their context, thus creating uncertainty as to the specific energy sources and technologies supported by the legislation. Where this occurred, an assessment was made based on a range of factors such as the context of usage throughout the legislation, whether the country was landlocked or not and the other energy sources included in the legislative definition.

It should be noted that of the 113 countries with renewable energy laws, 22 of those countries are landlocked.[175] As a result, these countries are less likely to support the ocean forms of renewable energy, though they may still support the riverine forms where they have access to such water sources. Interestingly, half of the landlocked countries such as Austria and Hungary do still include ocean energies within their legislative definition. These countries are all (with the exception of Malawi) European countries, with the vast majority being either EU Member States or EU candidate countries and thus are obliged to transpose the EU definition, which includes these energy sources. The inclusion of these energy sources, even though the countries are landlocked, may be used to facilitate the development of the internal European energy market, as well as a European trading scheme for renewable energy obligations in the future. This also reflects the EU's strategic plan to create prosperous conditions for the renewable ocean energy sector 'to enable the EU to capture a sizeable share of the market', which has been estimated to be worth up to €535 billion over the next forty years.[176]

171 Tester et al., above n 6, 597.

172 Ibid.

173 Judith Wallace, 'Tides, Waves, and Ocean Currents' in Michael B Gerrard (ed.), *The Law of Clean Energy: Efficiency and Renewables* (American Bar Association, Section of Environment, Energy and Resources, 2012) 509, 522.

174 Ibid 516–22.

175 The landlocked countries with national renewable energy laws are: Afghanistan, Andorra, Armenia, Austria, Belarus, Czech Republic, Hungary, Kosovo, Kyrgyz Republic, Liechtenstein, Luxembourg, Macedonia, FYR, Malawi, Moldova, Mongolia, Paraguay, San Marino, Serbia, Slovak Republic, Switzerland, Tajikistan, and Uzbekistan.

176 Commission of the European Communities, *Communication – Blue Energy – Action Needed to Deliver on the Potential of Ocean Energy in European Seas and Oceans by 2020 and Beyond* (COM (2014) 8 final), 20 January 2014.

In the case of Malawi, the inclusion of 'ocean energies' within their definition of 'renewable energy' is perhaps only explicable by the inappropriate transposition of a foreign definition of 'renewable energy' from a country with coastal regions.

2.7.1 *Tidal Power or Current Energy*

Tidal power takes advantage of the high and low tides created twice daily by the gravitational interactions between the earth/moon and sun.[177] The energy contained in the water caused by the variation in the height of the low and high tides or strong tidal currents in narrow passages is captured when the water flows through a turbine.[178] Tidal power stations use low-head, propeller-type turbines similar to the technologies used in freshwater run-of-the-river hydropower generation.[179] Tidal/current energy is included in the legislative definitions of renewable energy of seventy-five countries, as well as the EU and the IRENA definitions. It is excluded from the legislative definitions of twenty-nine countries. It was unclear whether tidal/current energy was included within the legislative definition adopted by nine countries; four countries had generally vague or ambiguous definitions, while five countries had elements of uncertainty in their definition which specifically affected the interpretation of tidal/current energy.

Tidal power is hampered by high initial capital costs, long construction times and very intermittent power generation.[180] A further limitation is that, for large-scale commercial generation to be viable, a tidal range of at least five metres is necessary.[181]

2.7.2 *Wave Energy*

Wave energy is derived by using a range of technologies to capture the hydrokinetic power of swells as they pitch, heave or surge.[182] Some of these technologies use the 'oscillating wave motions to create[183] hydraulic pistons, which force air through a turbine' to power an electric generator.

177 Tester et al., above n 6, 590.
178 Needham, above n 68, 11.
179 Tester et al., above n 6, 590.
180 Needham, above n 68, 11.
181 Ibid 11.
182 Tester et al., above n 6, 597; Needham, above n 68, 11.
183 Tester et al., above n 6, 597.

Simpler technologies involve using similar concepts to run of the river hydropower by inducing 'waves to run up ramps to replenish a low-head hydro reservoir'.[184] One significant limiting factor on the deployment of wave energy relates to the high power densities of waves during storms, which can cause serious damage to the generators.[185] Addressing this issue in the design of wave power generators is costly and requires a thorough understanding of the physics of complicated wave hydrodynamics.[186] Other issues relate to the high technology costs, complex licensing and consenting procedures, the lack of available finance and the current lack of subsea transmission infrastructure.[187]

Wave energy is included within the legislative definitions of renewable energy in sixty-nine countries, as well as the EU and the IRENA definitions. It is excluded from the definitions of thirty-six countries. The inclusion or exclusion of wave energy is uncertain in eight countries; four countries had generally vague or ambiguous definitions, while four countries had elements of uncertainty in their definition that specifically affected the interpretation of wave energy.

2.7.3 *Hydrothermal Energy (Maremotermica)*

Ocean thermal layering technologies have a limited geographic range, working most effectively between the tropical latitudes of 20° north and 20° south of the equator.[188] This is because in this zone, there is a temperature differential in the surface layers of the ocean of approximately 20–25°C to the deep water found several hundred metres below the surface.[189]

Ocean thermal energy conversion projects use the warm water near the surface to heat and then cool circulating fluids such as ammonia, or a mixture of ammonia and water.[190] The idea is that the warm surface water heats the circulating fluid in a vaporiser producing enough gas pressure to drive a turbine that in turn generates power. Once the gas has passed through the turbine, the circulating fluid is then condensed by passing it through pipes

184 Ibid.

185 Tester et al., above n 6, 597.

186 Ibid.

187 Commission of the European Communities, *Communication – Blue Energy – Action Needed to Deliver on the Potential of Ocean Energy in European Seas and Oceans by 2020 and Beyond* (COM (2014) 8 final), 20 January 2014.

188 Tester et al., above n 6, 599; Needham, above n 68, 14.

189 Ibid.

190 Needham, above n 68, 14. Other proposed circulating fluids include hydrogen and methane: Tester et al., above n 6, 599.

cooled by the cold deep water. The resulting fluid is recycled to be passed through the vaporiser and turbine again; the water that has acted as a coolant for the pipes is deposited back into the deep water and electricity is generated.[191]

Despite the great resource potential for thermal layering technologies, which has been estimated 'on the order of tens of thousands of Gigawatts',[192] there are several barriers to its large-scale deployment. Other than the geographic restriction required to ensure an adequate temperature differential, a further complication is that the presence of broad continental shelves around many countries means that it is either impossible to transmit the energy generated or significant load losses would be incurred throughout the transportation process.[193] Concerns have also been raised about the ability to generate energy efficiently when compared to gas turbine combined cycle generators given the high cost per kilowatt-hour (kWh) of thermal layering technologies,[194] as well as the higher operating and maintenance costs for seaborne technologies.[195]

Thermal layering technologies may provide some added benefits, such as the potential opportunity for mariculture brought about by bringing 'large amounts of cold, nutrient-rich seawater close to the surface'.[196] However, as the cold water also contains higher levels of dissolved carbon dioxide than warm water, the use of thermal layering technologies may release carbon dioxide into the atmosphere while generating energy.[197] That said, it is likely that if this is the case, thermal layering technologies would still release substantially lower carbon dioxide emissions than a fossil fuel combusting generator when generating the same amount of electricity.[198]

A final issue with thermal layering technologies is that countries need to be conscious of Article 56(1)(a) of the UN Convention on the Law of the Sea[199] which gives coastal States sovereign rights to the 'energy from the water,

191 See e.g. Tester et al., above n 6, 599 and Needham, above n 68, 14 for further details of this process.

192 Tester et al., above n 6, 599.

193 Ibid.

194 Ibid 602; see also Monique Hoogwijk and Wina Graus, *Global Potential of Renewable Energy Sources: A Literature Assessment* (Background Report, Ecofys, Renewable Energy Policy Network for the 21st Century, March 2008) 31.

195 Tester et al., above n 6, 602.

196 Ibid 606.

197 Ibid.

198 Ibid.

199 *United Nations Convention on the Law of the Sea*, opened for signature 10 December 1982, 1833 UNTS 3 (entered into force 16 November 1994).

currents and winds', within their exclusive economic zone. However, the exercise of these rights must be performed in a manner that is compatible with the other provisions of the Convention, including Article 87, which protects the freedom and safety of navigation, and Part XII, which governs the protection and preservation of the marine environment. Where more stringent regional regimes exist, these must also be complied with.[200]

One of the challenges in assessing the extent of support of hydrothermal energy within the legislative definitions of renewable energy is the inconsistent use of the term 'hydrothermal' amongst different countries. In many countries, 'hydrothermal' refers to ocean thermal energy or maremotermica. However, some twenty-nine countries[201] that are outside the geographic area in which ocean thermal energy is possible, predominantly the EU Member States, use the same term to refer to geothermal hot springs and aquifers.[202]

The inconsistent use of terminology without useful further explanatory definitions makes statutory interpretation difficult. For example, the EU currently defines hydrothermal energy as the 'energy stored in the form of heat in surface water', which could arguably describe both geothermal hot springs and maremotermica. However, members of the drafting team of the EU Renewable Energy Directive have subsequently made it clear that it was intended to only refer to geothermal hot springs.[203] This issue will be resolved by proposed changes to the recast EU Renewable Energy Directive contained in the Winter Package, which has removed the reference to 'hydrothermal energy' and replaced it with the phrase 'tidal, wave and other ocean energy'.[204]

As such, the number of countries with renewable energy laws that include 'hydrothermal energy' within their legislative definition is arguably less useful

200 Donald R Rothwell and Tim Stephens, *The International Law of the Sea* (Hart Publishing, 2010) 89.

201 These countries are: Albania, Argentina, The Bahamas, Bulgaria, Chile, Croatia, Cyprus, the Czech Republic, Egypt, Hungary, Italy, Kosovo, Latvia, Liechtenstein, Lithuania, Luxembourg, Macedonia, FYR, Moldova, The Netherlands, Norway, Pakistan, Poland, Portugal, Romania, Slovak Republic, Slovenia, Spain and Ukraine. Note that some other countries such as France, the United Kingdom and the United States whose mainland's are located outside of this geographic area are not included on this list due to the presence of overseas territories such as New Caledonia (France), Hawaii (USA) and Ascension (UK) within the area.

202 Niels Ladefoged, 'Definitions of Renewable Energy' in Paul Hodson, Christopher Jones and Hans Van Steen, *EU Energy Law: Renewable Energy Law and Policy in the European Union* (Claeys & Casteels, 2010) vol 3 book 1, 31, 32, 37–9.

203 Ibid.

204 General Secretariat of the Council of the European Union Proposal for a Directive of the European Parliament and of the Council on the promotion of the use of energy from renewable sources – Analysis of the final compromise text with a view to agreement, above n 66, 50, Preamble, recital 2(a).

than that for the other technologies. That said, fifty-one countries currently include some form of 'hydrothermal energy' within their legislative definition, as does the EU and the IRENA. Fifty-one countries exclude 'hydrothermal energy', with the inclusion or exclusion unclear in the legislative definitions of a further eleven countries.

2.7.4 *Osmotic Energy (Salt Gradient)*

While there is technology available to exploit salt gradients, including pressure-retarded osmosis, vapour compression and reversed dialysis technologies,[205] these technologies have not been subject to large-scale commercialisation. As a result, only two countries, Germany and Mexico, explicitly incorporate osmotic salt gradient energy into their legislative definition of 'renewable energy'.

Pressure-retarded osmosis works by establishing at the river mouth a freshwater reservoir (from the river) and a saltwater reservoir (made up of seawater) divided by a semi-permeable ion-specific membrane.[206] This membrane enables fresh water to pass through into the saltwater reservoir but prevents the completion of the osmotic process, which would normally equalise the salinity levels in both reservoirs by passing saltwater into the freshwater reservoir.[207] The increased pressure in the saltwater reservoir is then used to power a turbine generating electricity.

2.8 HYDROGEN/FUEL CELLS

Fuel cells produce electricity by converting the chemical energy of fuels, usually hydrogen derived from biogas or natural gas, and an oxidant (air or oxygen) into electricity.[208] There are five main types of fuel cells that each utilise different materials, have different processes and operating characteristics: proton exchange membrane fuel cells, alkaline fuel cells, phosphoric acid fuel cells, solid oxide fuel cells and molten carbonate fuel cells.[209]

205 Needham, above n 68, 15.

206 Needham, above n 68, 15; Dean Clark, 'Salt Power: Norway Project Gives Osmotic Energy a Shake', *National Geographic News* (online), 7 January 2013 <http://news.nationalgeographic.com/news/energy/2013/01/130107-osmotic-energy-norway/>.

207 Needham, above n 68, 15.

208 IEA, *Technology Roadmap: Hydrogen and Fuel Cells* (OECD/IEA, 2015), 8, 30–1.

209 Ibid 31. Charles Kubert, *Fuel Cell Technology: A Clean, Reliable Source of Stationary Power* (Clean Energy States Alliance, 2010), 2.

When hydrogen atoms enter the anode their molecules are separated to form protons, which are then passed through an electrolyte, and negatively charged electrons, which are passed through an external electrical circuit as direct current (DC). The protons and the electrons then recombine in the presence of oxygen at the cathode, to form water and heat, which is drained from the fuel cell.[210]

Fuel cells have very high levels of energy efficiency at levels of up to 70 per cent,[211] which is more than double that of conventional combustion technologies.[212] They can also provide a reliable form of baseload power (and heat), without emitting greenhouse gases or other air pollutants at the point of electricity generation.[213] The main limitation to the large-scale commercial exploitation of hydrogen fuel cells is that hydrogen 'cannot be mined or extracted in its elemental form. It instead needs to be separated from other compounds (such as water or hydrocarbon fuels)'.[214] This process not only requires energy but may also generate carbon dioxide as a by-product of the gas commonly used in the process.[215] Despite this, the carbon dioxide emissions are still lower than conventional fossil fuel generation.[216]

The question of whether hydrogen fuel cells are renewable must ultimately turn on the question of whether the gas used to separate out the hydrogen comes from a renewable source such as landfill gas, sewage treatment gas or biogas, or from natural gas, which is a fossil fuel.[217] However, this continues to be a field of substantial research and development, meaning that it should eventually be possible to derive the hydrogen through the process of water electrolysis. Five countries currently include hydrogen and/or fuel cells within their legislative definition of renewable energy: Bangladesh, Ireland, Paraguay, Pakistan and Tonga.

210 See e.g. Paul Breeze, 'Fuel Cells' in Breeze, above n 143, 131–4; Charles Kubert, *Fuel Cell Technology: A Clean, Reliable Source of Stationary Power* (CESA, 2010), 1 <www.cesa.org/assets/Uploads/Resources-pre-8-16/CESA-fuelcelltechnology-may2010.pdf>.

211 IEA, *Technology Roadmap: Hydrogen and Fuel Cells*, above n 208, 30.

212 Office of Energy Efficiency & Renewable Energy, *Fuel Cells Basics* (1 June 2014) <www.energy.gov/eere/fuelcells/fuel-cell-basics>.

213 Ibid.

214 Charles Kubert and Warren Leon, 'Hydrogen Production and Storage' in Clean Energy States Alliance (ed.), *Fuel Cells: Briefing Papers for State Policymakers* (CESA, 2011) 11.

215 Breeze, above n 143, 129.

216 Kubert and Leon, above n 214, 129.

217 See e.g. Ruven Fleming and Joshua P Fershee, 'Hydrogen Economy, in the United States and the European Union' in Donald Zillman, Lee Godden, LeRoy Paddock and Martha Roddenkamp (eds.), *Innovation in Energy Law and Technology: Dynamic Solutions for Energy Transitions* (Oxford University Press, 2018), 148.

2.9 PEAT

Peat is defined in the draft International Recommendations Energy Statistics Manual prepared by the UN Statistical Commission as 'a solid formed from the partial decomposition of dead vegetation under conditions of high humidity and limited air access (initial stage of coalification) and any products derived from it'.[218] This means that peat is formed from biomass that, due to its constant regeneration, is deemed to be a renewable resource. This argument formed the basis of Finland's argument that peat should be included in the definition of renewable energy adopted by the EU in the 2009 Renewable Energy Directive.[219] Finland, with the largest remaining peatland in Europe, stood to gain significantly if peat was included in the definition. However, despite its organic origins, all of the major international organisations, including the EU, and every country with a national renewable energy law (except Sweden) categorise peat as a fossil fuel because it takes the right environmental conditions and thousands of years to renew, thus making it more akin to a fossil fuel such as coal or lignite. Furthermore, the greenhouse gas emissions produced by the burning of peat, which is the method used to generate electricity from the main peat products, namely sod peat, milled peat and peat briquettes, are similar to those of fossil fuels when conducted on a life cycle analysis.[220] There is only one country that has included peat within their renewable energy legislation. The Swedish Electricity Certificates Act (2011) acknowledged that peat is not a renewable energy source, with it falling outside of their legislative definition. However on closer inspection of the Swedish legislation, there is a supplementary definition for 'renewable electricity' that is defined as 'electricity produced from renewable energy sources or peat'.[221] This suggests that while Sweden acknowledges that peat is in fact not a renewable energy source as it is depleted at a faster rate than it can be replaced through natural processes, it still wants to support an indigenous energy source.

218 UN Statistical Commission, *Draft International Recommendations on Energy Statistics* (United Nations, 2011) 38–9.

219 Directive 2009/28/EC of the European Parliament and of the Council of 23 April 2009 on the promotion of the use of energy from renewable sources and amending and subsequently repealing Directives 2001/77/EC and 2003/30/EC [1993] OJ L 140/16.

220 Kiyoto Tanabe, 'Fuel classification and definitions in the 2006 IPCC National Greenhouse Gas Inventory Programme Guidelines' (Paper presented at the 2nd InterEnerStatWorkshop, Paris, 19–20 November 2007), 6 <www.iea.org/work/2007/interenerstat/IPCC.pdf>.

221 *Lag om elcertifikat* [Electricity Certificates Act] (Sweden) No. 2011:120, s 2[1–2].

2.10 NUCLEAR ENERGY

Nuclear reactors generate electricity and heat by converting the energy released from the nucleus of an atom using the process of nuclear fission. The heat generated from this process is used to generate steam, which in turn drives a turbine connected to a generator.

The first regulatory support for renewable energy in the United Kingdom was the Non-Fossil Fuel Obligation Order (NFFO), which was established on 1 October 1990 to provide financial support to the operators of nuclear power stations. The original purpose of the NFFO was to provide a subsidy to the State-owned nuclear companies, following the privatisation of the rest of the electricity generation sector in the United Kingdom in 1989.[222] The NFFO later came to be used by the renewable energy sector, prior to the introduction of the Renewables Obligation (RO) Order in April 2002.[223]

Ever since the introduction of regulatory support for the accelerated deployment of renewable energy, there have been repeated attempts to label nuclear energy, both domestically within the United Kingdom and the United States,[224] and internationally in arenas such as the EU and the IRENA as a renewable resource.[225] If nuclear power were accepted as a renewable energy source, the consequences for other renewable energy sources could be quite significant. For example, in England and Wales, it would enable nuclear generators to be exempted from the Climate Change Levy. The Climate Change Levy is a tax borne by the fossil fuel and nuclear industry despite the Intergovernmental Panel on Climate Change (IPCC) describing the emissions generated by nuclear power as follows:

222 Catherine Mitchell, 'The Renewables NFFO: A Review' (1995) 23 *Energy Policy* 1077.

223 Catherine Mitchell and Peter Connor, 'Renewable Energy Policy in the UK 1990–2003' (2004) 32 *Energy Policy* 1935.

224 House of Commons Innovation, Universities, Science and Skills Committee, *Renewable Electricity-generation Technologies*, House of Commons Report No. 5, Session 2007–08 (2008); David Elliott, 'Sustainable Energy: Nuclear Power and Renewables' in David Elliott (ed.), *Sustainable Energy: Opportunities and Limitations*, Energy, Climate and Environment Series (Palgrave MacMillan, 2007) xviii.

225 Lily Riahi, 'Promise of New Renewable Energy Agency in Peril?', *The Huffington Post* (online), 19 July 2009 <www.huffingtonpost.com/lily-riahi/promise-of-new-renewable_b_217335.html>; David Gow and Ian Traynor, 'Nuclear Question Splits EU Climate Talks', *The Guardian* (online), 8 March 2007 <www.theguardian.com/environment/2007/mar/08/eu.europeanunion>; Terry Macalister, 'Nuclear Industry Accused of Hijacking Clean Energy Forum', *The Guardian* (online), 28 June 2009 <www.theguardian.com/business/2009/jun/28/nuclear-industry-global-body-plans>; World Nuclear News, 'EU Sets Carbon and Renewables Targets, Backs Nuclear', *World Nuclear News* (online), 9 March 2007 <www.world-nuclear-news.org/newsarticle.aspx?id=13028>.

> The life cycle GHG emissions per kWh from nuclear power plants are two orders of magnitude lower than fossil-fuelled electricity generation and comparable to most renewables.[226]

It would also enable nuclear generators to enter into long-term supply contracts for renewable generation, with guarantees of purchase for the electricity generated, to the detriment of new renewable generators.[227] It is for this reason that the nuclear industry has engaged in extensive lobbying to have nuclear energy labelled as renewable.[228]

2.10.1 *Are Nuclear Feedstocks a Renewable Resource?*

Nuclear power generation makes use of nuclear feedstocks such as uranium-235 and plutonium-238. These feedstocks have to be mined from the earth and have a finite lifespan and, as such, are not renewable. However, if a similar approach to geothermal energy is adopted in relation to nuclear power, then the argument of physicist Bernard Cohen in 1983 that uranium is effectively inexhaustible and could therefore be considered a renewable source of energy[229] might give the argument that nuclear power is a renewable resource some credence.

However, when this issue was debated in the United Kingdom House of Lords in 2005, the Science and Innovation Minister, Lord Sainsbury of Turville, stated that it was important not to characterise the meaning of renewable energy within the context of the debate as a finite or indefinite source of energy but rather:

> The technical question of whether nuclear energy is or is not a renewable source of energy turns on how one defines 'a renewable source of energy', and the view one takes on the supply of uranium, the use of other materials and the commercial prospects of nuclear fusion.[230]

226 Intergovernmental Panel on Climate Change (2001) quoted in Elliott, above n 224, 7.

227 Carl Mortished, 'Minister Declares Nuclear "Renewable"', *Times* (London), 31 October 2005 <www.timesonline.co.uk/tol/business/industry_sectors/utilities/articles584657.ece>.

228 Jean Eaglesham and Christopher Adams, '"Green" Subsidy Considered for Nuclear Power', *Financial Times* (London), 26 October 2005 <www.ft.com/cms/s/0/a423189c-45bc-11da-981b00000e2511c8.html>.

229 Bernard L Cohen, 'Breeder Reactors: A Renewable Energy Source' (1983) 51 *American Journal of Physics* 78.

230 United Kingdom, *Parliamentary Debates*, House of Lords, 22 November 2005, House of Lords, Debate on Nuclear Energy, Hansard, 22 November 2005, <www.publications.parliament.uk/pa/ld200506/ldhansrd/vo051122/text/51122-02.htm>; Ibid.

2.10.2 *Do the Other Environmental Benefits of Nuclear Generation Outweigh the Consequences of Its Use and Therefore Justify Its Inclusion Within the Definition of 'Renewable Energy'?*

Due to the very low greenhouse gas emissions produced over the whole life of nuclear power stations,[231] which are only slightly higher than most renewable resources, the argument has been made that nuclear energy should be incorporated into the definition of 'renewable energy'. For example, in 2005 in the United Kingdom, the then Energy Minister Malcolm Wicks argued that significant benefits would be derived from extending the Renewables Obligation to include 'nuclear and other low carbon sources of energy, such as clean coal'.[232] However, Elliott has argued that a more appropriate measure for assessing environmental impacts would

> consider wider ecological principles, such as the idea that we should avoid introducing major irreversible changes. For example, taking a long-term perspective, it could be argued that the release of radioactive materials, some with active lifetimes of several thousand years, represents a serious and irreversible burden to the ecosystem, which should be avoided at all costs as a matter of principle. By contrast, most renewable energy technologies have relatively small environmental impacts and, if necessary, the equipment can be removed leaving at most only very minor short-term damage.[233]

Verbruggen and Yurchenko have also argued that

> nuclear power plants fail on all sustainability criteria, except for being low-carbon ... The significant risks and the permanence of the nuclear waste bequest create an irreversible burden, contradictory to the essence of sustainable energy transitions.[234]

For this reason, although nuclear power has repeatedly been recognised as a low-carbon energy source, the efforts to label nuclear energy as a renewable energy source have been almost universally resisted. Ecuador is the only country that has recognised nuclear energy as a renewable energy source:

231 A Adamantiades and Ioannis Kessides, 'Nuclear Power for Sustainable Development: Current Status and Future Prospects' (2009) 37 *Energy Policy* 5149, 5150–1.

232 Eaglesham and Adams, above n 228.

233 Elliott, above n 224, 13.

234 Aviel Verbruggen and Yuliya Yurchenko, 'Positioning Nuclear Power in the Low-Carbon Electricity Transition' (2017) 9 *Sustainability* 163.

> 'Renewable energy': is energy derived from sources that do not diminish from being utilized: hydropower, wind, solar, geothermal, biomass, tidal, nuclear and others.[235]

This reflects Ecuador's desire to develop nuclear energy generation capacity. These ambitions have also been highlighted through Ecuador's 2009 Nuclear Cooperation Agreement with Rosatom, a Russian State Nuclear Energy Corporation,[236] and recent statements before the UN General Assembly and the International Atomic Energy Agency.[237]

The American state of Utah also includes nuclear generation within the definition of 'renewable energy' contained in its Renewable Energy Development Act 2009:

> s63 M-1–2803(6): 'renewable energy' ... includes generation powered by nuclear fuel.

No other country, nor the EU Directive or the IRENA Statue have ever categorised nuclear power as 'renewable energy'. Indeed, it is far more common for countries explicitly to exclude nuclear energy from their definition of renewable energy than include it, with Hungary, South Korea and the United Kingdom just a few of the countries to adopt this approach. For example, s 32 of the Electricity Act 1989 (England, Wales and Scotland) explicitly states that:

> 'renewable sources' means sources of energy other than fossil fuel or nuclear fuel.

2.10.3 *Why Is Nuclear Energy Non-Renewable When Geothermal Energy Is a Recognised Source of Renewable Energy?*

On the one hand, the exclusion of nuclear power from the definition of 'renewable energy' seems appropriate due to the finite nature of uranium, the negative environmental externalities and health and safety risks potentially associated with use of nuclear power. On the other hand, it is difficult to justify

235 *Ley Orgánica del Servicio Público de Energía Eléctrica* [Organic Law on the Public Service of Electricity, Electricity Law 2015], Art 3 [Linguistico Translations translation from Spanish] (Ecuador).

236 World Nuclear Association, *Emerging Nuclear Energy Countries* (September 2018) <www.world-nuclear.org/information-library/country-profiles/others/emerging-nuclear-energy-countries.aspx>.

237 United Nations, 'Nuclear Energy Could Hold Key to Sustainable Development Gains, Delegates Tell General Assembly, as It Considers International Atomic Energy Agency Report, Meetings Coverage of the Seventy-Second Session, 46th and 47th Meetings, GA/11972', 10 November 2017, <www.un.org/press/en/2017/ga11972.doc.htm>.

nuclear energy not being renewable based on the depletion rate argument if we accept geothermal energy as renewable because it depletes at such a slow rate as to be essentially inexhaustible. The justification for nuclear energy not being renewable, while geothermal energy is renewable, seems to be in the conflation of the definition of renewable energy with an unstated requirement that the energy also be sustainable. Elliot has argued that the terms 'renewable', 'sustainable' and 'green energy' are often used interchangeably and 'sometimes used simply to imply that the source has good environmental credentials',[238] while Carleton has noted that 'renewable resources are considered sustainable ipso facto'.[239] However, all energy sources used to generate electricity have both positive and negative impacts both on the environment, in terms of health and safety, and economically. It is inconsistencies such as these that create confusion in the minds of ordinary people about what 'renewable energy' is, especially when the requirement that renewable energy also be sustainable is most often an unwritten implied rule.

2.11 ANY OTHER SOURCE PRESCRIBED BY THE REGULATIONS OR OTHERWISE PERMITTED

Fourteen countries explicitly permit other sources of energy to be prescribed as a renewable energy source either through the amendment of regulations, decree by the Minister or a decision of the relevant energy authority. Forty-one countries do not permit such an option. It is unclear whether such a move would be permitted or prohibited using this method as opposed to legislative amendment in the remaining fifty-eight countries with national renewable energy laws.

2.12 CONCLUSION

The definitions of 'renewable energy' used in legislation often reflect the historical origins of the regulatory support for renewable energy in each jurisdiction, and the outcomes of political debate about the strategic direction of sub-sectors of the energy industry. In many instances, the definition of renewable energy adopted or advocated by different jurisdictions is reflective of their indigenous energy sources (both renewable and non-renewable). Thus

[238] David Elliott, 'Introduction: Sustainable Energy: The Options' in David Elliott (ed.), *Sustainable Energy: Opportunities and Limitations* (Palgrave Macmillan, 2007) xviii.

[239] Alexandra L Carleton, 'Mandating Market Access for Renewable Energies in Australia' (2008) 26 *Journal of Energy and Natural Resources Law* 402, 403.

each jurisdiction's definition is the product of their unique social and environmental context.

Despite this, there was a surprising degree of consensus in the major energy sources identified as renewable within the definitions. Indeed at least 80 of the 113 countries with national renewable energy laws accept each of the following energy sources as renewable: wind energy, solar energy (both photovoltaic and concentrated solar thermal), biomass, landfill gas, sewage treatment gas and biogas, small-scale hydropower and geothermal energy. This level of conceptual consensus indicates that differences to the form and the approach to the content of the legislative definition of renewable energy often do not make a material difference for the most commercialised renewable energy sources other than to impose transaction costs and act as an additional market barrier for new market entrants. Currently the level of additional costs that this imposes on renewable electricity companies operating across international borders is not known and thus further research by economists is required on this point. However, it does suggest that there is an opportunity for greater efficiencies to be achieved within the renewable energy sector through the movement towards a standard approach to the form and content of the legislative definition of renewable energy internationally. This conceptual consensus as to the subject matter of the national renewable energy laws of different countries may provide the basis for later international harmonisation or legislative convergence in years to come.

However, not all potential renewable energy sources benefit from this level of conceptual consensus within the legislative definitions of renewable energy. Most notably, large-scale hydropower was only accepted as a renewable energy source in 77 countries, as compared to the 102 countries that accept small-scale hydropower. The levels of acceptance were even lower for the ocean renewable energy sources such as wave, hydrothermal (maremotermica) and osmotic (salt gradient) energy. Even when the twenty-two landlocked countries that had national renewable energy laws were excluded from consideration, ocean energy is still significantly less likely to receive support than many other sources of scientifically recognised renewable energy. This may be attributed to their lower levels of commercialisation, which may act as a barrier to their immediate support in developing countries that require an 'off the shelf' technological solution due to the lack of funds for research and development. The risk profile for projects using emerging technologies for generating ocean and marine energy is also materially higher than projects using the more accepted energy sources, making them unsuitable for developing countries unless they have an exceptional comparative advantage in their resource.

The energy sources that were least likely to be accepted as renewable energy were those that arguably would not be accepted within the scientific meaning of renewable energy (i.e. energy which is depleted at an equal or slower rate than its replacement rate). The inclusion of these energy sources within the definition of renewable energy increases uncertainty in the meaning of the term and exemplifies the presence of over-inclusive legislative definitions. In the future, hydrogen fuel cells are likely to be a source of renewable generation and as a result, it is likely that over time they will garner broader acceptance within the legislative definition of renewable energy. However, the feedstocks currently used as sources of hydrogen are commonly fossil fuel hydrocarbons and thus, even though hydrogen fuel cells have lower carbon emissions than conventional fossil fuel generation, current levels of acceptance are low. Meanwhile, peat, charcoal and nuclear energy, all energy sources whose status as 'renewable' or 'non-renewable' are often debated, were only accepted within the legislative definitions of a single country each.

The analysis of the current debates about whether particular energy sources are renewable has highlighted that many of the decisions as to whether a type of energy should be deemed to be renewable or not, are not based exclusively or predominantly on scientific considerations or consistent legal principles. Rather, they highlight the contested and inherently political nature of the process of defining 'renewable energy'. How else can we explain why there are debates about peat being a renewable energy source when it is the first stage of coalification, when coal has never been considered to be renewable? Or justify geothermal energy being defined as renewable because it depletes at such a slow rate as to be essentially inexhaustible, when the same arguments can be made for nuclear energy, which has been repeatedly found to be a non-renewable source of energy? On the basis of this research, the answer seems to be in the conflation of the definition of renewable energy with an unstated requirement that the energy also be sustainable and preferably highly commercialised.

PART II

WHY DO COUNTRIES INTERVENE IN THE RENEWABLE ENERGY SECTOR? A CASE OF NORMATIVE DIVERGENCE

3

The Economic Justification for Regulating Renewable Energy

A majority of the countries in the world now engage in some form of government intervention within the renewable energy sector, with the numbers consistently growing year on year. By 2018, 146 countries had renewable power targets,[1] 138 countries had support policies directed at renewable energy[2] and 113 countries had national renewable energy laws in force. In this context, it is important to understand the economic rationale for intervening in the renewable energy sector, as well as the factors driving this growth in the development of national renewable energy laws.

3.1 THE CHARACTERISTICS OF ELECTRICITY THAT WARRANT SPECIAL REGULATORY TREATMENT

Electricity is a secondary source of energy and an essential component of the modern economy. It has traditionally been generated from fossil fuels such as coal, natural gas and oil, though an increasing quantity is now generated from renewable and nuclear sources.

Economists and regulatory theorists define electricity as a 'mixed good' that possesses characteristics of both a private good that would be best regulated by market forces, and a public good that would be best regulated by public policy.[3]

1 REN21 Secretariat, 'Renewables 2018 Global Status Report' (Report, Renewable Energy Policy Network for the 21st Century, 2018) Table R8.

2 Ibid.

3 Mark Jaccard, 'Oscillating Currents: The Changing Rationale for Government Intervention in the Electricity Industry' (1995) 23 *Energy Policy* 579, 582; William Blyth and Kirsty Hamilton, *Aligning Climate and Energy Policy: Creating Incentives to Invest in Low Carbon Technologies in the Context of Linked Markets for Fossil Fuel, Electricity and Carbon* (Chatham House, 2006) 1.

Similar to a private good, electricity is rivalrous[4] and excludable,[5] but it also provides non-rivalrous and non-excludable public benefits for which preferences are not effectively revealed by market mechanisms. The characteristics that support the definition of electricity as a 'mixed good' warranting regulatory intervention include:

1. electricity is essential for individual welfare and economic development;
2. electricity generation is interdependent with other fuel sources;
3. electricity pricing does not accurately reflect the true cost of electricity generation due to the presence of externalities and information asymmetries;
4. demand for electricity is uneven and largely unresponsive to short-term price spikes;
5. electricity cannot currently be economically stored in most countries;[6] and
6. there is historically significant market concentration in the electricity sector.

There is considerable variation in national electricity markets because of the different endowments of indigenous fossil fuels and renewable energy potential,[7] levels of technological development and environmental and energy security risks. As a result, the relative importance of each of the above characteristics defining electricity as a 'mixed good' varies between countries. In this section, each of the above characteristics will be assessed to highlight why energy markets have an increased likelihood of market failure without regulatory intervention.

3.1.1 *Electricity Is Essential for Economic Development and Individual Welfare*

Global demand for electricity is constantly increasing due to economic growth, rising populations and more energy-intensive projects. Electricity is a key component of the modern economy, contributing to economic development through its critical role in industrial production, transport and

4 A good is considered rivalrous where its consumption by one individual reduces its availability to another individual.

5 A good is considered excludable when the owner can prevent other individuals from consuming it.

6 Though this is rapidly changing as cost reductions occur in energy storage technologies.

7 See e.g. Richard L Ottinger, Lily Mathews and Nadia Elizabeth Czachor, 'Renewable Energy National Legislation: Challenges and Opportunities' in Donald N Zillman et al. (eds.), *Beyond the Carbon Economy: Energy Law in Transition* (Oxford University Press, 2008) 183, 184–5; Commonwealth Productivity Commission, 'Carbon Emission Policies in Key Economies' (Research Report, Productivity Commission, 2011) xvi; Commonwealth Department of Resources, Energy and Tourism, 'National Energy Security Assessment 2009' (Assessment, Commonwealth of Australia, 2009) 1.

telecommunications, and contributing to social wellbeing through its use in refrigeration, cooking, heating and entertainment.[8] Indeed, Blyth and Hamilton have argued that such is the importance of electricity that:

> in OECD countries, access to electricity has almost come to be seen as a 'right' – expectations of high levels of reliability of electricity supply means that the electricity system is highly politicised. This puts electricity uncomfortably on the boundary between a private good suited to control by market forces and a public good suited to control by public policy.[9]

This idea of access to energy being a human right is also reflected in the UN's Sustainable Development Goals, with Goal 7 being to 'ensure access to affordable, reliable, sustainable and modern energy for all'.[10]

Due to the essential nature of electricity to society, there are a number of stakeholders actively interested in the organisation and regulation of the energy sector. They include the different levels of government, industrial and domestic consumers, employees, electricity distribution and supply companies, corporate shareholders and the regulators. These stakeholders possess different interests that need to be balanced in long-term and strategic national energy policies.[11] However, many countries have failed to implement a coordinated and integrated national energy policy and an accompanying regulatory regime. Where this is the case, governments may struggle to ensure that energy production and use provide wider public benefits that are in the national interest, such as environmental sustainability, energy security and meeting growing energy demands. Indeed, Adetoro has argued that the combination of these factors and the fact that the electricity sector concerns not merely business but also directly has an impact on the social welfare and living standards of citizens, means that it is an area likely to be highly politicised.[12] Thus the nature of electricity as a 'mixed good' makes electricity vulnerable to political interference and a target of lobbying activity.

[8] Bastian Becker and Doris Fischer, 'Promoting Renewable Electricity Generation in Emerging Economies' (2013) 56 *Energy Policy* 446, 446; Commonwealth Department of Resources, Energy and Tourism, above n 7, 1. See also Emmanuela Colombo, Lorenzo Mattarolo and Stefano Mandelli, 'Global Dimension of Universal Access to Energy' in Emmanuela Colombo, Stefano Bologna and Diego Masera (eds.), *Renewable Energy for Unleashing Sustainable Development* (Springer, 2013), 27.

[9] Blyth and Hamilton, above n 3, 1.

[10] United Nations, *Sustainable Development Goals* (2018) <https://sustainabledevelopment.un.org/>.

[11] David Adetoro, 'Can the Imposition of a Regulator in Any Liberalised Energy Market Be Justified by Market Behaviours?' (2006) 24 *Journal of Energy & Natural Resources Law* 384, 387.

[12] Adetoro, above n 11, 387.

3.1.2 *Electricity Generation Is Interdependent with Other Fuel Sources*

Due to the reliance on primary energy sources in generation, 'changes in demand for electricity and relative price shifts between the fuels can therefore lead to complex feedback between the prices of each energy source because of the linkage through power generation'.[13] This feedback mechanism is further complicated by the fact that each primary energy source has an uneven global geographic distribution and has its own political and regulatory dynamics that drive its consumption, which vary considerably by region and country.[14] This makes the energy sector inherently political and uncertain. For example, the world's proven conventional oil and gas reserves are highly concentrated, with members of the Organization for the Petroleum Exporting Countries (OPEC) accounting for 81.9 per cent of global conventional oil reserves.[15] In comparison, the OECD countries that consume close to 50 per cent of the world's oil account for only 14.3 per cent of global proved oil reserves.[16] However, electricity production and consumption is not only affected by geopolitical volatility affecting primary fossil fuel sources; it may also be affected by significant policy shifts in major markets or accidents. For example, renewable energy received a significant boost in the EU under the Large Combustion Plant Directive, which required that large fossil fuel generators built prior to 1987 must either comply with emissions standards or close by 2015.[17] The 2011 Fukushima nuclear disaster in Japan also led to a number of countries, including Italy,[18] Germany[19] and Switzerland,[20] imposing moratoria on the construction of new nuclear generation facilities or the extension of life for existing nuclear facilities. Other factors that have had an impact on the choice

13 Blyth and Hamilton, above n 3, 1.

14 Ibid.

15 OPEC, OPEC Share of world crude oil reserves 2017 <www.opec.org/opec_web/en/data_graphs/330.htm> (accessed 10 August 2018).

16 IEA, *Oil Market Report 2018: Analysis and Forecasts to 2023* (OECD/IEA, 2018), 28, 33; BP, *BP Statistical Review of World Energy* (67th edn, BP, 2018) 12 <www.bp.com/content/dam/bp/en/corporate/pdf/energy-economics/statistical-review/bp-stats-review-2018-full-report.pdf>.

17 *Directive 2001/80/EC of the European Parliament and of the Council on the Limitation of Emissions of Certain Pollutants into the Air from Large Combustion Plants* [2001] OJ L 309/1.

18 World Nuclear News, 'Italy Announces Nuclear Moratorium', *World Nuclear Association* (online), 24 March 2011 <www.world-nuclear-news.org/NP-Italy_announces_nuclear_moratorium-2403117.html>.

19 Spiegel Online International, 'Germany to Reconsider Nuclear Policy: Merkel Sets Three-Month "Moratorium" on Extension of Lifespans', *Spiegel Online* (online), 14 March 2011 <www.spiegel.de/international/world/0,1518,750916,00.html>.

20 Judy Dempsey, 'Nuclear Plants in Europe Are Delayed', *The New York Times* (online), 14 March 2011 <www.nytimes.com/2011/03/15/business/global/15euronuke.html?_r=2&src=busln>.

of fuels used in the electricity generation market are the plant replacement cycles for aged conventional fossil fuel and nuclear generators,[21] as well as energy security concerns rising out of the Russia/Ukraine conflict and civil unrest in Nigeria, Iraq and Libya.

3.1.3 *Electricity Pricing Does Not Accurately Reflect the True Cost of Generation Due to Externalities and Information Asymmetries*

Electricity pricing does not accurately reflect the economic and societal cost of electricity generation due to the presence of externalities and information asymmetries. There are significant social and environmental externalities associated with electricity generation, with 'externality' best understood as 'a negative or positive effect of some activity that is experienced by a third party, but is not accounted for in an associated purchase transaction or payment for damage'.[22] For example, the pricing of coal-fired electricity generation fails adequately to price the externalities associated with its use including air pollution, high greenhouse gas emissions and health impacts such as higher rates of asthma, lung cancer and heart disease suffered in communities surrounding coal-fired generation plants.[23] The issue of greenhouse gas emissions is particularly acute, as electricity generation accounts for 45 per cent of global energy-related carbon dioxide emissions.[24] As the costs of these negative externalities are not incorporated into electricity prices, unless an emissions trading scheme is in place, the market will encourage a level of fossil fuel generation that is economically inefficient and sub-optimal for society as a whole.[25]

Electricity pricing not only neglects to price negative externalities but it also fails to ascribe an appropriate value to positive externalities such as the

21 See e.g. World Nuclear News, 'Last Decade of German Nuclear Power', *World Nuclear Association* (online), 31 May 2011 <www.world-nuclear-news.org/newsarticle.aspx?id=30132&terms=german%20moratorium>.

22 Jaccard, above n 3, 582.

23 Shruti Khadka Mishra, *Estimation of Externality Costs of Electricity Generation from Coal: An OH-Markal Extension* (PhD Thesis, The Ohio State University, 2009) 6; Ottinger et al., above n 7, 188; Philippe Menanteau, Dominique Finon and Marie-Laure Lamy, 'Prices Versus Quantities: Choosing Policies for Promoting the Development of Renewable Energy' (2003) 31 *Energy Policy* 799, 800–1; Mike Sandiford et al., Submission to the Productivity Commission, *Productivity Commissions' Draft Report on Electricity Network Regulation Public Inquiry*, 2013, 7 <www.pc.gov.au/inquiries/completed/electricity/submissions/subdr073-electricity.pdf>.

24 IEA and IRENA, *Perspectives for the Energy Transition: Investment Needs for a Low-Carbon Energy System* (OECD/IEA and IRENA, 2017) 39.

25 Jaccard, above n 3, 582.

benefit of increased energy security derived from having a diversified energy supply.[26] This is because while large industrial electricity consumers may have sufficient information to price the loss of a fuel source due to market or physical disruption, the average domestic consumer or small or even medium-sized business owner does not have access to the same resources.[27] As a result, only large industrial consumers can make informed decisions as to the cost of taking preventative and/or remedial steps to ensure they can use diverse fuel sources. The presence of information asymmetries in the electricity sector has led Adetoro to state that 'policy makers in the energy industries have to contend with ... a monopoly in the supply of information'.[28]

Clement-Davies et al. have proposed that cost efficiency projections over the life of the project should be undertaken for all new energy projects to ensure that renewable generation, with its higher upfront capital costs, is evaluated on a more even playing field.[29] Such a move would include the initial capital costs, feedstock costs, maintenance costs and decommissioning costs. However, given the volatility in fossil fuel commodities prices and the absence of an enforceable global regulatory regime for countries that fail to meet their Nationally Determined Contributions under the Paris Agreement, these models can be unreliable. Indeed, studies have shown that models of fossil fuel generation often fail to adequately account for price rises in the feedstock over time or alternatively, assume that there will be no price rise at all.[30] Further, many financial models used within the sector still fail to ascribe a value to externalities.

26 Trent Berry and Mark Jaccard, 'The Renewable Portfolio Standard: Design Considerations and an Implementation Survey' (2001) 29 *Energy Policy* 263, 264; Molly F Sherlock and Donald J Marples, 'Energy Tax Policy: Issues in the 111th Congress' (Report, Congressional Research Service, 20 September 2010) 3 <http://nationalaglawcenter.org/wp-content/uploads/assets/crs/R40999.pdf>; Severin Borenstein, 'The Private and Public Economies of Renewable Electricity Generation' 26 *The Journal of Economic Perspectives* 67; Kenneth Gillingham and James Sweeney, 'Market Failure and the Structure of Externalities' in Boaz Mozelle, Jorge Padilla and Richard Schmalensee (eds.), *Harnessing Renewable Energy* (RFF Press, 2010) 69, 72.

27 Lionel Chin, Ross Gawler and Walter Gerardi, 'NEM Market Failures and Governance Barriers for New Technologies' (Final Report to Garnaut Climate Change Review, McLennan Magasanik Associates, 2008) 15–16.

28 Adetoro, above n 11, 398.

29 Christopher Clement-Davies et al., 'Renewables Investment' (2009) 6 *International Energy Law Review* 213, 213–14.

30 Energy Infrastructure Assurance Advisory Group and National Oil Supplies Emergency Committee, *Diesel Fuel & Back-Up Generation: Issues for CEOs, Risk Managers and Diesel Users* (Trusted Information Sharing Network, 2009).

3.1.4 *Demand for Electricity Is Uneven and Largely Unresponsive to Short-Term Price Spikes*

Demand for electricity fluctuates according to a range of variables such as the weather, the time of year and the economic cycle. Further, in the traditional model of transmission and distribution networks that is still used by most countries, an individual consumer can elect to vary the amount of power they draw from the system at any given time with no notice.[31] Research across many jurisdictions has also indicated that in the absence of large-scale deployment of smart meters and cost-reflective pricing, consumer demand tends to be largely unresponsive to short-term price spikes in the spot market.[32] A key challenge in the electricity sector is the need to balance demand with supply. Without the installation of smart grids and meters, there is no way for a consumer to receive up-to-date information about the amount or cost of power they consume. As a result, consumers often choose to use power at a time that is most convenient for them, which is likely to coincide with the peak demand times for other users.[33] While smart grids have already been deployed in some countries, a substantial majority of the world's population have still not adopted this technology and thus this is likely to continue being a problem for some time to come.

This limited consumer response to price variation is evidence of an inefficient electricity market leading to market failure and potential energy insecurity. The IEA has argued that this 'limited demand response is a legacy of old vertically integrated energy systems, where the focus was largely set on the supply side and energy prices were uniformly set'.[34] However, when buyers do not participate actively in the price-setting process, prices cannot play their balancing role leading to excessive price volatility. This has prompted the IEA to advocate better consumer demand response, which 'can notably dampen price peaks, reducing costs and risks to all market participants'.[35]

3.1.5 *Electricity Cannot Currently Be Economically Stored in Most Countries*

The widespread adoption of residential, commercial and grid-scale energy storage systems is likely to be one of the most transformative changes to the

31 Blyth and Hamilton, above n 3, 1–2.

32 See e.g. Ahmad Faruqui, Sanem Sergici and Lamine Akaba, 'Dynamic Pricing of Electricity for Residential Customers: The Evidence from Michigan' (2013) 6 *Energy Efficiency* 571.

33 Ibid.

34 IEA, *Energy Security and Climate Policy: Assessing Interactions* (OECD/IEA, 2007) 35; see also IEA, *Empowering Customer Choice in Electricity Markets* (OECD/IEA, 2011) 8.

35 Ibid.

energy sector in the coming decade. While there have been significant technological improvements in energy storage systems, in many countries it is still not cost-competitive for electricity consumers (whether at a residential, commercial or utility scale) to store their energy.[36] This means that when a consumer demands electricity, supply across the transmission and distribution networks must be carried out in real time. This is beginning to change in some countries. These countries fall within two categories, those that have generous regulatory support mechanisms and those that have plentiful renewable resources, high electricity prices and low or non-existent feed-in tariffs.[37] Australia falls into the latter category and has seen the rapid uptake of battery storage systems, with a tripling of residential installations to an estimated 20,800 in 2017,[38] even in the absence of subsidies. However, until other countries reach this point, the intermittent and variable nature of renewable energy presents a problem for generators, suppliers and regulators who have to ensure that there is sufficient supply to meet an uncertain and constantly changing level of demand. The need to respond to unexpected demand peaks means that there needs to be a mix of generation fuel sources used within the energy sector to ensure that generation can be ramped up quickly if necessary. This 'ramp up' capacity is currently often met through the use of fossil fuel sources such as coal- or gas-fired generation or large-scale hydropower, as the primary inputs for this electricity can be easily stored and the output is not intermittent.[39]

3.1.6 *There Is Historically Significant Market Concentration in the Electricity Sector*

A further challenge within the electricity sector is that there is significant market concentration in the sector, with a small number of large generating companies using highly capital-intensive and long-lived assets. This reflects the fact that, historically, the electricity generation sector in many countries has had high barriers to entry, with approvals, financing and access to technology and infrastructure difficult for market entrants to obtain. Previously, these high barriers to entry meant that it was not economically viable to build

36 IRENA, *Adapting Renewable Energy Policies to Dynamic Market Conditions* (IRENA, 2014) 12.

37 REN21 Secretariat, above n 1, 159.

38 Ibid.

39 IEA, *Energy Security and Climate Policy: Assessing Interactions*, above n 34, 52; see also IRENA, *Planning for the Renewable Future: Long-Term Modelling and Tools to Expand Variable Renewable Power in Emerging Economies* (IRENA, 2017) 33.

generation plants with a capacity of less than 500MW.[40] Furthermore, shortages in proven technology and difficulties gaining planning permission acted to prevent new market entrants from entering the electricity generation sector.[41]

Mitchell and Woodman have argued that:

> . . . conditions in the selection environment tend to support larger companies, with access to cheaper financing of economies of scale, rather than smaller companies which have less access. A side-effect of a technology and fuel-blind market is that it is less supportive of new, smaller, diverse entrants, which might undertake new and innovative activities. New and smaller companies may be more flexible or may be able to survive by finding niches for themselves by doing different activities (such as providing energy services; installing a solar water heater and so on). Larger companies are unlikely to undertake new and innovative activities because in order to stay competitive they are more likely to concentrate on doing the same activities, but more cheaply.[42]

These high barriers to entry and the selection conditions for large companies mean that the electricity generation market has traditionally been highly concentrated.[43] Indeed, until recently in many countries, electricity was provided by national monopoly utility companies.[44] These high barriers to entry and the degree of market concentration may limit competition in the electricity sector, enabling the incumbents to exercise market power. Anti-competitive behaviour can 'lead to over-charging and restriction of the supply of goods and services relative to the social optimum'.[45]

3.2 MARKET FAILURES AFFECTING THE RENEWABLE ENERGY SECTOR

The market is generally viewed to be the most efficient and effective mechanism for ensuring the allocative efficiency of goods and

40 Adetoro, above n 11, 398.

41 Fredric Beck and Eric Martinot, 'Renewable Energy Policies and Barriers' in Cutler Cleveland and Robert Ayers (eds.), *Encyclopedia of Energy (Volume 5)* (Elsevier, 2004) 365, 367.

42 Catherine Mitchell and Bridget Woodman, 'Towards Trust in Regulation – Moving to a Public Value Regulation' (2010) 38 *Energy Policy* 2644, 2648.

43 Chin et al., above n 27, 4.

44 Andras Lakatos, 'Overview of the Regulatory Environment for Trade and Electricity' in Janusz Bielecki and Melaku Geboye Desta (eds.), *Electricity Trade in Europe: Review of the Economic and Regulatory Challenges* (Kluwer Law International, 2004) 120.

45 Chin et al., above n 27, 4.

services.[46] However, the characteristics of electricity identified above that make it a 'mixed good' also make it vulnerable to market failures and the imposition of market barriers that may 'inhibit socially optimal levels of investment'.[47]

Brown has defined market failures as the 'conditions of the market that violate one or more of the neoclassical economic assumptions that define an ideal market for products or services such as rational behaviour, cost of transactions, and perfect information'.[48] There are multiple sources of market failure present in the electricity sector. However, 'identifying market failures, assessing their magnitude and designing adequate policies is not a straightforward task'.[49]

While there is debate between economists as to which market failures exist in the electricity sector, there appears to be broad consensus for the presence of at least three different market failures. The first market failure is the failure to price the negative externalities associated with fossil fuels such as the health and environmental impacts into the fossil fuel price, and conversely the failure to price the positive externalities associated with renewable energy, such as its role in mitigating climate change and ensuring energy security into the renewable energy price. The second market failure relates to the effects of positive spillovers and learning effects. This means that private firms engaging in research and technological innovation in the sector may not receive the full returns on their investment due to some of those benefits being publicly shared (or capable of being reversed engineered). The third market failure is the presence of information asymmetries resulting from hidden characteristics, principally in the form of uncertainty around future market developments and future generation costs for both fossil fuels and renewable energy.

3.2.1 *Positive Spillovers and Learning Effects*

One of the most significant market failures within the renewable energy sector is the impact of spillovers and learning effects on research and development.

46 Cabinet Office Performance and Innovation Unit, *Public Services: The Rationale for Government Intervention* (Government of the United Kingdom, 2001) <http://webarchive.nationalarchives.gov.uk/+/http://www.cabinetoffice.gov.uk/strategy/downloads/survivalguide/downloads/rationale.pdf>; Katrin Jordan-Korte, *Government Promotion of Renewable Energy Technologies: Policy Approaches and Market Development in Germany, the United States, and Japan* (Gabler Research, 2011) 22; Barry Barton et al. (eds.), *Energy Security: Managing Risk in a Dynamic Legal and Regulatory Environment* (Oxford University Press, 2004) 4.

47 Marilyn A Brown, 'Market Failures and Barriers as a Basis for Clean Energy Policies' (2001) 29 *Energy Policy* 1197, 1199.

48 Ibid.

49 IEA, *Energy Security and Climate Policy: Assessing Interactions*, above n 34, 3.

Spillovers occur where knowledge generated through the research and development process or subsequent commercialisation by one firm spills over to benefit other firms.[50] This means that firms and other sectors of the economy that have not contributed time and money to the development of the knowledge may still be able to benefit from that knowledge despite not compensating the original innovating firm. This knowledge transfer may occur through the original innovating firm not being able to protect their intellectual property (i.e. such as through competitors reverse engineering technological innovations), communication between firms and participation in industry events such as conferences, as well as staff movements between firms.[51] For this reason, spillovers are characterised from an economic perspective as a positive externality.

Learning effects (also sometimes referred to as 'learning-by-doing' or 'economies of experience'[52]) describes the process by which production costs decline as experience increases. As with the market failures associated with research and development, spillovers are also created through the initial process of adoption and diffusion of a new technology. This again means that the first firm to adopt new technologies may be 'unable to appropriate the complete social returns of their knowledge'.[53]

According to economic theory, the presence of positive spillovers and learning effects leads to 'invention and innovation failures, caused by underinvestment in basic research and development'.[54] This is an example of dynamic inefficiency[55] because it discourages the first moving firm from investing more in research and development than the amount that they can appropriate as private returns.[56] There is a lack of agreement amongst the academic literature both about the extent and impact of positive spillovers and learning effects on the renewable energy sector and about the appropriate

50 Paul Lehmann and Erik Gawel, 'Why Should Support Schemes for Renewable Electricity Complement the EU Emissions Trading Scheme?' (Discussion Paper, Helmholtz Department of Economics, 2011) 4; Aaron Cosbey, 'Green Industrial Policy and the World Trading System' (Issue Brief, Entwined, 2013) <www.iisd.org/publications/green-industrial-policy-and-world-trading-system> 4; Sandiford, above n 23, 5.

51 Lehmann and Gawel, above n 50, 4.

52 Severin Borenstein, 'Government Subsidies for Renewable Energy: When Is Pricing Greenhouse Gases Not Enough?' (Speech delivered at the Haas Policy Conference, Sacramento, 26 October 2009) 7.

53 Lehmann and Gawel, above n 50, 4.

54 Ottmar Edenhofer et al., 'On the Economics of Renewable Energy Sources' (2013) 40(1) *Energy Economics* S12, S18.

55 Borenstein, 'The Private and Public Economies of Renewable Electricity Generation', above n 26, 82.

56 Lehmann and Gawel, above n 50, 4.

response to their presence.[57] One of the difficulties in assessing the extent of the impact of learning effects in particular is that it is often difficult to separate those effects from other changes within the sector,[58] such as currency fluctuations or changes to the labour market making technologies less expensive, or increasing economies of scale.

Anecdotally, there does seem to be a strong argument that spillovers relating to both research and development and learning effects are having an impact on the sector. In particular there have been a number of allegations of reverse engineering and intellectual property law violations within the sector, often relating to Chinese manufacturers.[59]

3.2.2 *The Presence of Unpriced Negative and Positive Externalities*

One of the central precepts of welfare economics is that individual and social welfare may be maximised through the optimal allocation of resources.[60] As stated above, one of the most significant economic issues affecting the optimal allocation of resources in the energy sector is the presence of unpriced negative and positive externalities. It can be challenging to price an externality as it 'involves identifying the nature of the external impacts, identifying the parties affected, and estimating implied costs'.[61] The IEA has argued that this process is difficult because:

> '1. impacts can be widely diffused and exhaustively identifying all parties affected may be impossible;
> 2. estimating costs requires distinguishing the externality from other market imperfections which can also be difficult; and
> 3. external impacts can involve considerations related to health, the environment or equity, which can be difficult to evaluate in monetary terms'.[62]

57 Lehmann and Gawel, above n 50, 4–6; Sandiford, above n 23, 5; Brown, above n 47, 1201–2; Borenstein, 'The Private and Public Economies of Renewable Electricity Generation', above n 26, 82–3; Edenhofer et al., above n 54, S18–20.

58 See e.g. Borenstein, 'The Private and Public Economies of Renewable Electricity Generation', above n 26, 82–3.

59 Peter Behr, 'Chinese Company Accused of Economic Espionage in Wind Turbine Case', *E&E News* (online), 26 January 2012 <www.eenews.net/stories/1059958987>.

60 Beatriz Yordi, *Communication on Support Schemes for Electricity from Renewable Energy Sources: External Costs of Energy and Their Internalisation in Europe* (ExternE, 2006) 1; Gillingham and Sweeney, above n 26, 5.

61 IEA, *Energy Security and Climate Policy: Assessing Interactions*, above n 34, 3.

62 Ibid.

Further, in order to be able to make a sensible comparison of different energy sources, there needs to be a consistent model used to internalise externality costs across all of the competing energy forms or technologies.[63] A number of attempts have been made to estimate the costs of externalities in electricity production. In a study by Biegler, which used the ExternE method of pricing externalities, the unpriced externalities relating to greenhouse gas emissions and health damage associated with electricity generation in Australia were $AUD19/MWh for natural gas, $AUD42/MWh for black coal and $AUD52/MWh for brown coal.[64] This may be compared with the average wholesale price for electricity in the Australian National Electricity Market at the time of the study of $AUD40/MWh.[65] This same study found that the unpriced externalities that were associated with renewable generation were low, with estimates provided of $AUD5/MWh for photovoltaic solar generation and $AUD1.50/MWh for onshore wind generation.[66] These costs were largely due to the embedded emissions derived from the manufacture of these technologies.[67] The cumulative impact of these unpriced externalities, and the associated market impact, is felt globally. Indeed, the IRENA have priced 'the external effects of energy supply and use related to climate change and air pollution . . . in the order of USD 2.2 trillion–USD 5.9 trillion per year'.[68] This urgent need to address the inequity resulting from unpriced externalities forms the basis of the argument that there should be national regulatory intervention in the electricity sector.

3.2.3 *Information Asymmetries*

In perfectly competitive and efficient markets, all market participants have free access to perfect information so that rational decisions may be made based on informed choices.[69] Ryan et al. have stated that 'information has characteristics which resemble that of a pure public good: it is non-rivalrous in

63 Jaccard, above n 3.

64 Tom Biegler, *The Hidden Costs of Electricity: Externalities of Power Generation in Australia'* (The Australian Academy of Technological Sciences and Engineering, 2009), ii.

65 Ibid i.

66 Ibid ii.

67 Ibid.

68 IRENA, *The True Cost of Fossil Fuels: Saving on the Externalities of Air Pollution and Climate Change* (IRENA, 2016) 6.

69 For further information about the economic theory behind imperfect information see Joseph E Stiglitz and Bruce C Greenwald, 'Externalities in Economies with Imperfect Information and Incomplete Markets' (1986) 101(2) *Quarterly Journal of Economics* 229; and see generally the work of George Akerlof, Michael Spence and Joseph E Stiglitz.

consumption (its use by one person does not reduce its availability to another person) and non-excludable in ownership (it is difficult for the entity producing information to exclude other people from its benefits or to capture all the benefits of its use)'.[70]

Information asymmetries occur where the relevant information is either not freely available or where insufficient or poor quality information is provided to some market participants. There are commonly two different situations in which information asymmetries may inhibit the effective functioning of a competitive market: adverse selection; and moral hazard. Adverse selection describes the situation where one party in a transaction (commonly the seller) has access to superior information about a good than the other party (commonly the buyer).[71] In contrast, moral hazard 'refers to a situation in which one party in a transaction's actions are unobservable to the other, leading the first party to act opportunistically after a contract for buying or selling a good or service is signed'.[72] Of these two situations, adverse selection is more likely to pose a problem for market participants in the renewable energy sector.

Within the renewable energy sector, Sawin[73] and Mirza et al.[74] have identified a range of information failures, which may increase the perceived risks of investing in renewable energy:

1. a lack of information or misconceptions about available renewable energy resources;
2. a lack of information or misconceptions about the current state of renewable energy technologies;
3. high costs and barriers associated with accessing the relevant information;
4. a lack of experience or training regarding renewable energy projects; and
5. a lack of understanding about the full range of costs, benefits and risks associated with different energy sources, technologies and projects.

These information failures may lead market participants to make sub-optimal decisions, which in turn can affect the competiveness of the market and investment decisions. In particular, where market participants cannot easily

70 Lisa Ryan, Sara Moarif, Ellina Levina and Richard Baron, 'Energy Efficiency Policy and Carbon Pricing' (Information Paper, OECD/IEA, 2011) 13.

71 Ibid 14–15; Jordan-Korte, above n 46, 29.

72 Ryan et al., above n 70, 15.

73 Janet Sawin, 'National Policy Instruments: Policy Lessons for the Advancement & Diffusion of Renewable Energy Technologies Around the World' (Paper presented at the International Conference for Renewable Energies, Bonn, 2004) 1–2.

74 Umar K Mirza, Nasir Ahmad, Khanji Harijan and Tariq Majeed, 'Identifying and Addressing Barriers to Renewable Energy Development in Pakistan' (2009) 13 *Renewable and Sustainable Energy Reviews* 927, 929.

access information about the costs of renewable energy relative to fossil fuel generation, they are likely to remain with their incumbent energy source.[75] This is likely to have a negative impact on the deployment of renewable energy.

As stated above, one of the challenges faced by new market entrants is that there are pervasive barriers to entry, such as problems accessing relevant information and high capital costs within the renewable energy sector. This was reflected by the commentary from the IEA that, within the context of the Chinese power sector:

> relevant information is not always made public. Thus, few stakeholders understand current pricing methods and the complex fee structures. It is also hard to judge current levels of profit, subsidies and cross subsidisation. Opportunities to participate in the regulatory process are unclear.[76]

These barriers impose significant costs on new market entrants who must try to negotiate complex and highly technical information and procedures. New market entrants may be further constrained by the lack of transparency and accountability sometimes present in the sector, often in the form of ad hoc decision-making. This creates additional expense and places administrative burdens on new market entrants who may not be privy to information essential to make informed decisions on likely market developments. It also decreases their responsiveness to developments in the sector, as they may have insufficient information.

Government policy and regulation must provide more support to new market entrants. New market entrants would benefit from the provision of accurate and timely information by government regulators. Further, regulatory or policy specialists should provide guidance in respect of the procedures needed to gain approvals and the steps followed by relevant decision-makers. Where the market is unwilling to invest, or the cost of capital makes borrowing unaffordable, one-off low interest loans to assist with the greater capital costs involved in renewable energy start-ups may also be of benefit.

3.3 MARKET BARRIERS WITHIN THE RENEWABLE ENERGY SECTOR

Further to the causes of market failure in the electricity sector, obstacles in the form of market barriers may also be present and contribute to the slow

75 Ottinger et al., above n 7, 185; Menanteau et al., above n 23, 801–2.

76 IEA, *China's Power Sector Reforms: Where to Next?* (OECD/IEA, 2006) 72.

diffusion and adoption of innovations in the electricity sector.[77] The market barriers present in the renewable electricity sector include:

1. subsidies to fossil fuel sources and their use in electricity generation;
2. regulatory uncertainty and fragmented policy-making;
3. barriers to access to the transmission and distribution networks;
4. planning permission and approvals;
5. economies of scale of the projects within the renewable energy sector;
6. limited access to finance and capital market failures;
7. principal-agent problems/split incentives;
8. lack of skilled labour;
9. lack of social/consumer acceptance; and
10. imperfect competition.

Much of the previous research has characterised these barriers into five categories: policy barriers, institutional barriers, economic barriers, technological barriers and infrastructure barriers.[78]

3.3.1 *Subsidies to Fossil Fuels and Their Use in Electricity Generation*

One of the most significant barriers to the accelerated deployment of renewable energy is the subsidies provided to fossil fuels and nuclear generation. These subsidies take a number of forms, with research on fossil fuel subsidies in OECD countries identifying twenty-four different forms of subsidies available.[79] Typically, the subsidies offered to fossil fuels are either designed to drive demand in the form of consumer subsidies (where the price paid by consumers is beneath a benchmark price) or drive supply in the form of producer subsidies (where the prices received by suppliers are above a benchmark price).[80]

77 See e.g. Brown, above n 47, 1199.

78 Catherine Mitchell, *Energy, Climate and Environment Series: The Political Economy of Sustainable Energy* (Palgrave Macmillan, 2010) 68; Jyoti Prasad Painuly, 'Barriers to Renewable Energy Penetration: A Framework for Analysis' (2001) 24 *Renewable Energy* 73, 75; Corinna Klessmann, Anne Held, Max Rathmann and Mario Ragwitz, 'Status and Perspectives of Renewable Energy Policy and Deployment in the European Union – What Is Needed to Reach the 2020 Targets?' (2011) 39 *Energy Policy* 7637, 7651–2.

79 Benjamin K Sovacool, *Renewable Electricity for Southeast Asia: Designing the Right Policy Architecture* (Lee Kuan Yew School of Public Policy, National University of Singapore, 2009) 9.

80 IMF, *Energy Subsidy Reform: Lessons and Implications* (IMF, 2013) 6 <www.imf.org/en/Publications/Policy-Papers/Issues/2016/12/31/Energy-Subsidy-Reform-Lessons-and-Implications-PP4741>.

When not appropriately limited and targeted, subsidies can 'aggravate fiscal imbalances, crowd-out priority public spending, and depress private investment'.[81] Further, subsidies are known to cause serious problems with the efficient allocation of resources within the energy sector by encouraging overconsumption of energy, altering the incentives for investment in renewable energy, and accelerating the depletion rates of finite natural resources.[82] The IEA has estimated that $US260 billion was provided in direct subsidies to fossil fuel (excluding nuclear generation) in 2016 alone[83] (or 0.34 per cent of global GDP).[84] This may be compared to the direct subsidies provided to renewable energy used for power generation over the same period of $US140 billion.[85]

While these subsidies are now a sunk cost, they have created a powerful historical legacy of technological lock-in in favour of fossil fuel energy sources and generating technologies. Sovacool has noted that conventional sources (i.e. fossil fuels and nuclear energy sources) 'received almost 90 per cent of all subsidies for the past six decades'.[86] Thus, one of the most important things that governments can do to support electricity generation from renewable sources is to remove the subsidies they provide to fossil fuel and nuclear generation. In 2016, the G7 pledged to phase out inefficient fossil fuel subsidies by 2025,[87] while in July 2017, the G20 affirmed their intent to phase out 'inefficient fossil fuel subsidies that encourage wasteful consumption over the medium term'.[88] This issue also features in Agenda 2030 for Sustainable Development, which was agreed by 193 Member States of the United Nations:

> Rationalize inefficient fossil-fuel subsidies that encourage wasteful consumption by removing market distortions, in accordance with national circumstances, including by restructuring taxation and phasing out those harmful subsidies, where they exist, to reflect their environmental impacts, taking fully into account the specific needs and conditions of developing countries

81 Ibid 1.

82 Ibid 5; Kirsty Hamilton, 'Scaling Up Renewable Energy in Developing Countries: Finance and Investment Perspectives' (Paper No. 02/10, Chatham House, 2010) 5.

83 IEA, *World Energy Outlook* (OECD/IEA, 2017) 85.

84 Author's own calculation using the global gross domestic product data for 2016 from World Bank, *World Development Indicators database* (1 July 2018) <https://databank.worldbank.org/data/download/GDP.pdf>.

85 IEA, *World Energy Outlook*, above n 83, 82.

86 Sovacool, above n 79, 1533.

87 G7, *Ise-Shima Leaders' Declaration G7, Ise-Shima Summit*, 26–27 May 2016, 28 <www.mofa.go.jp/files/000160266.pdf>.

88 G20, *Annex to G20 Leaders Declaration, G20 Hamburg Action Plan*, Hamburg, Germany, 7–8 July 2017, 15 <www.g20.org/profiles/g20/modules/custom/g20_beverly/img/timeline/Germany/2017-g20-hamburg-action-plan-en.pdf>.

> and minimizing the possible adverse impacts on their development in a manner that protects the poor and the affected communities.[89]

If these reforms are realised, the renewable energy sector will not be the only beneficiary. The IMF believes that the removal of fossil fuel subsidies would generate positive spillover effects by reducing global energy demand. This would lead to an estimated reduction of global energy related CO_2 emissions of more than 20 per cent.[90] It is also likely to generate health benefits leading to a 55 per cent reduction in global air pollution deaths[91] through reducing sulphur dioxide and nitrous oxide emissions and other local pollutants such as fine particulate matter. However, this was not the first time that these organisations have pledged to remove inefficient fossil fuel subsidies, with both the G7 (then the G8) and the G20 having made similar pledges in 2009,[92] albeit without an explicit deadline, which have not been realised. While these subsidies continue to exist, a vicious cycle of market distortions operates whereby renewable energy must be subsidised or subject to other forms of government intervention to enable it to compete with fossil fuels.

3.3.2 *Regulatory Uncertainty and Fragmented Policy-Making*

In order to encourage investment and reduce perceived market risk, emerging markets such as those associated with the renewable energy sector need certain, stable and coordinated policies and regulations. However, due to its highly political nature, the energy sector is often prone to sudden shifts in policy, with policy measures typically following 'an erratic process of political decision-making which is driven by a variety of short-term concerns and considerations'.[93] Research conducted by Chatham House identified that three areas act as particular impediments to investors in the renewable energy sector:

> political risk in a country; policy and regulatory risk (certainty, visibility and a degree of immunity from political change); and capital or financial risk.

89 *Transforming Our World: The 2030 Agenda for Sustainable Development*, GA Res 70/1, 70th sess, Agenda Items 15 and 116, UN Doc A/RES/70/1 (21 October 2015, adopted 25 September 2015) <-https://sustainabledevelopment.un.org/post2015/transformingourworld>.

90 David Coady et al., 'How Large Are Global Energy Subsidies?' (IMF Working Paper, WP/15/105, May 2015) 26.

91 Ibid.

92 Jocelyn Timperley, *Explainer: The Challenge of Defining Fossil Fuel Subsidies* (12 June 2017) Carbon Brief <www.carbonbrief.org/explainer-the-challenge-of-defining-fossil-fuel-subsidies>.

93 Lehmann and Gawel, above n 50, 10.

> The priority for governments should be tackling policy and regulatory issues, as financiers already have experience in understanding and managing the other two.[94]

There are numerous examples of regulatory uncertainty and short-term decision-making having an impact on the renewable energy sector. For example, in the months following the Fukushima nuclear disaster, a number of countries cancelled their long-term stated policies to prolong the operating life of their nuclear power plants.[95] Another example, which had an immediate impact on the renewable energy sector, was the introduction of the carbon price in Australia on 1 July 2012 and then, following the election of a new government, its subsequent repeal on 17 July 2014. The introduction of the carbon price made investing in renewable energy more attractive by altering the price signals to the market about the energy sources and technologies in which there should be investment. Following its introduction, companies investing in the renewable energy sector would have factored the existence of the carbon price into their financial models for projects, which may last up to thirty years. When the carbon price was repealed, not only did this change the financial model for existing projects in the sector, but it also meant that a risk premium needed to be factored in for all future projects as the regulatory and policy environment was uncertain. When coupled with the introduction of more stringent planning permissions for onshore wind projects, this led to a significant drop-off in the number of new large-scale projects within the sector.[96] This is a common problem in the renewable energy sector with Fabrizio finding that 'firms invested less in new assets in states that had previously passed and repealed legislation to restructure the electricity industry, indicating that perceived regulatory instability reduces new investment and undermines policy goals'.[97]

The IEA has argued that the choice of policy instrument can also be problematic, with the most effective instruments, such as taxes, often being politically unpopular.[98] In some countries, this has led to greater levels of market intervention than would otherwise be undertaken, which can lead to

94 Hamilton, above n 82, 19.

95 Lehmann and Gawel, above n 50, 10.

96 Peter Hannam, 'Australia's Large-scale Renewable Investment Dives in 2014', *Sydney Morning Herald* (online), 12 January 2015 <www.smh.com.au/environment/climate-change/australias-largescale-renewable-investment-dives-in-2014-20150112-12mbis.html>.

97 Kira R Fabrizio, 'The Effect of Regulatory Uncertainty on Investment: Evidence from Renewable Energy Generation' (2013) 29 *Journal of Law, Economics and Organization* 765, 765.

98 IEA, *Energy Security and Climate Policy: Assessing Interactions*, above n 34, 3.

market distortion.[99] This happened in the State of New South Wales (NSW) in Australia, where a solar rebate was provided at $AU0.60 cents per kWh generated.[100] This programme was budgeted at $AU335 million, however, due to a cost blowout the estimated cost prior to the reduction of the benefit was more than $AU1.9 billion.[101] This led to the rapid growth of solar energy in NSW and when the scheme was ended early, led to compensation claims from the industry.[102]

A further problem that is evident in a number of jurisdictions made up of a federation of states is a lack of federal coordination, meaning that different rules have emerged in different states. Similar problems arise between different levels of government regulating on the same topic. Theoretically, this is known as regulatory competition and means that investment should move to the states with the most optimal regulatory approach and that regulation should harmonise over the medium to longer-term. However, prior to the point of harmonisation, a lack of coordination adds significant transaction costs and makes it more difficult to operate nationally.[103] For example, a non-exhaustive stocktake of emissions reductions policies by the Australian Productivity Commission found that in 2011 the United States had 307 different policies affecting emissions reductions.[104] The same study found that Australia had 237 different policies, Germany had 131, the United Kingdom had 104, China had 82, South Korea had 69, India had 68 and Japan had 67 different policies.[105] The Productivity Commission stated that the 'sheer numbers of these policies say little in themselves about the materiality or effectiveness of the aggregate response made by governments'.[106] However, it is also evident that there is much overlap and inconsistency in the regulation and policy mix in many countries,[107] and that these complexities have only been

99 See e.g. Beck and Martinot, above n 41, 367; Blyth and Hamilton, above n 3, 4–5; Commonwealth Department of Resources, Energy and Tourism, 'Energy White Paper – National Energy Policy Framework 2030' (Strategic Directions Paper, Department of Resources, Energy and Tourism, 2009) 10.

100 Australian Broadcasting Corporation, 'Fraser Changes Tune on Solar Rebate Cuts', *ABC Mid North Coast NSW* (online), 25 May 2011 <www.abc.net.au/news/stories/2011/05/25/3226563.htm?site=midnorthcoast>.

101 Ibid.

102 Australian Broadcasting Corporation, 'O'Farrell's Solar Bonus Stance Melts Under Pressure', *ABC News* (online), 24 May 2011 <www.abc.net.au/news/2011-05-24/ofarrells-solar-bonus-stance-melts-under-pressure/2728654>.

103 Mirza et al., above n 74, 929.

104 Commonwealth Productivity Commission, above n 7, 15.

105 Commonwealth Productivity Commission, above n 7, 15.

106 Ibid 147.

107 Ibid.

added to in recent years. 'Different levels of government can be supporting the same project, not adding to abatement but adding to cost (for example, the US Federal Government subsidies for renewable energy and state mandatory renewable energy targets)'.[108] Indeed, the Productivity Commission Report even found that overlaps existed within the same level of government.[109] This creates unnecessary complexity, high administration and transaction costs and is particularly damaging to new market entrants. If a new market entrant has to understand and determine how to comply with a large number of policies, it can be very costly and may potentially undermine their willingness to invest. Such an approach particularly disadvantages states with smaller markets (and thus smaller economies of scale), as the presence of these transaction costs may mean that it is not economically viable for companies to participate in these markets.

These barriers to new entry in the electricity sector have a particular impact on renewable generation, which tends to be smaller-scale and with less proven technologies.[110] The problems associated with market failures and market barriers are often addressed through regulatory intervention. However, as noted above, access to a reliable and affordable source of electricity is often treated as a right and many stakeholders, each expressing different interests, are involved, making the regulation of renewable energy inherently political. There is a risk that regulatory agencies subject to daily political pressures may place more weight on short-term political goals than a long-term and strategic approach to the regulation of the national energy sector.[111] If regulators focus on short-term political objectives rather than a long-term strategy, there is a risk that investors will not believe the sector is sufficiently stable to encourage appropriate levels of investment and that political rather than national interests will be served.[112]

3.3.3 *Barriers to Access to the Transmission and Distribution Networks*

Large-scale renewable energy projects are often only viable if they can transport their electricity to end-users via the transmission and distribution networks. Without grid access, renewable energy projects cannot sell the electricity they generate and their assets become stranded and unable to

108 Ibid.

109 Ibid.

110 Beck and Martinot, above n 41, 366.

111 Mark A Jamison and Sanford V Berg, *Annotated Reading List for a Body of Knowledge on Infrastructure Regulation* (The World Bank, 2008) 4.

112 Ibid.

realise the income required to repay their project loans. As a consequence, where renewable energy projects are unable to connect to the grid easily and in a timely manner, this is a significant market barrier. Ottinger et al. have identified a number of features of this barrier including:

1. unreasonable interconnection requirements;
2. excessive standby rates;
3. high transmission access rates;
4. excessively burdensome approval requirements for the interconnection of intermittent resources; and
5. availability commitments (i.e. performance or payment guarantees out of proportion with the generator's size or not recognising the intermittent nature of the renewable resource).[113]

A further complicating factor is that a number of different stakeholders may impose this barrier. Some governments may be reluctant to change their current generation mix due to historical technological lock-in[114] and vested political interests, such as the ongoing government ownership of key transmission and distribution assets. Grid network operators may be resistant to change because balancing customer demand with electricity supply is more difficult to predict and manage with intermittent generating sources,[115] or because, in some instances, the network operator also holds conventional generation assets.

For example, in the mid-2000s renewable generators in the United Kingdom were required to 'invest then connect' to the transmission and distribution networks.[116] This meant that more than 15GW of renewable generation in the United Kingdom was unable to connect to the transmission network.[117] Due to the location of the best renewable resources in the United Kingdom, and hence new renewable projects, there has been 'unprecedented

113 Ottinger et al., above n 7, 189–90.

114 Cosbey, above n 50, 5.

115 Union of Concerned Scientists, *Barriers to Renewable Energy Technologies, Powerful Solutions: Seven Ways to Switch America to Renewable Electricity* (UCS, 1999) 21–2. This process is, however, getting easier over time as more data are collected, making the modelling of generation and demand easier to predict: Mirza et al., above n 74, 929.

116 Frontier Economics, 'An Assessment of the Potential Impact on Consumers of Connect and Manage Access Proposals' (Report, United Kingdom Office of Gas and Electricity Markets, 2009) 7.

117 Scottish and Southern Energy, 'Memorandum Submitted by Scottish and Southern Energy: The Future of Britain's Electricity Networks' (Report, Energy and Climate Change Select Committee of the UK Parliament, 2009) <www.publications.parliament.uk/pa/cm200910/cmselect/cmenergy/194/194we43.htm>.

demand for network capacity in areas of the system with relatively little transmission'.[118] Thus, generators were typically offered connection to the transmission network contingent on future network reinforcements and/or the installation of new infrastructure. However, before these future network reinforcements and/or the installation of new infrastructure can take place, planning and other consenting processes need to be completed.[119] More than half the generation in 2009 in the transmission queue had been offered a contingent connection date after 2015.[120] This posed significant issues for renewable generators who were unable to recoup their high initial capital costs or utilise their stranded infrastructure until they were connected to the transmission network. Following a Transmission Access Review, these problems led to the 'invest then connect' approach being dropped in favour of a 'connect then manage' approach within the United Kingdom so that renewable energy generators could be connected to the grid in a timely manner.[121] While this has greatly assisted the problems experienced in the United Kingdom, these problems continue to be experienced in other countries with less sophisticated infrastructure and limited resources.

3.3.4 *Planning Permission and Approvals*

Long lead times have been associated with the grant of planning permission in the renewable energy sector.[122] The determination of planning permission applications also tends to be the area of greatest political interest and interference within the renewable energy sector. For example, the planning process for the Beauly-Denny transmission line in Scotland, which was urgently required to enable renewable generation projects access to the grid, was considered for over six years by the authorities.[123] This prompted the Brown

[118] Mark Copley, 'The GB Queue – Problems and Possible Solutions' (Paper, United Kingdom Office of Gas and Electricity Markets, 2007) 4. This problem is not unique to the UK, and has been reported in other jurisdictions: Mirza et al., above n 74, 929.

[119] Copley, above n 118, 4.

[120] Scottish and Southern Energy, above n 117.

[121] Department of Environment and Climate Change and Ofgem, *Electricity network delivery and access* (11 January 2013) <www.gov.uk/electricity-network-delivery-and-access>.

[122] See e.g. IRENA, *Overcoming Barriers to Authorizing Renewable Power Plants and Infrastructure* (IRENA, 2013) 7–8; Nigel Martin and John Rice, 'Improving Australia's Renewable Energy Project Policy and Planning: A Multiple Stakeholder Analysis' (2015) 84 *Energy Policy* 128–41; Hamilton, above n 82, 20.

[123] Kristy Dorsey, 'Beauly-Denny: Shock to the System', *Scotland on Sunday (Edinburgh)* (online), 10 January 2010 <www.scotsman.com/business/beauly-denny-shock-to-the-system-1-1363889>.

Labour Government to put in place National Planning Statements, including one specifically on renewable energy, and to establish the Infrastructure Planning Commission to deal with planning determinations for critical national infrastructure.[124]

The development of onshore wind projects has been particularly affected by problems associated with planning permission.[125] These problems include 'siting and development approvals, due to the lack of familiarity with such installations on the part of officials and authorities, complicated planning approvals, and the misgivings of the public'.[126] These issues have been particularly noticeable in Japan,[127] the Netherlands and Australia.[128] In the latter case, there have been repeated public inquiries into the planning regime for wind farms,[129] with concerns being expressed about the need to adopt the precautionary principle in respect to Wind Turbine Syndrome, despite repeated public studies both in Australia and overseas concluding that there was no evidence that such a syndrome existed.[130] Other features of the planning regime for onshore wind in the Australian states, such as having some of the largest setbacks from residential developments and towns[131] and a noise standard that is 10 decibels (dB) lower than the European standard,[132] have all had a negative impact on the ability to develop new wind projects in Australia.[133]

124 Ibid.

125 Geraint Ellis et al., 'Wind Power: Is There a "Planning Problem"? Expanding Wind Power: A Problem of Planning, or of Perception? The Problems of Planning – A Developer's Perspective Wind Farms: More Respectful and Open Debate Needed, Not Less Planning: Problem "Carrier" Or Problem "Source"? "Innovative" Wind Power Planning' (2009) 10 *Planning Theory & Practice* 521.

126 Ottinger et al., above n 7, 190.

127 Emi Mizuno, 'Overview of Wind Energy Policy and Development in Japan' (2014) 40 *Renewable and Sustainable Energy Reviews* 999, 1006–10, 1016–17.

128 Ibid.

129 New South Wales Department of Planning and Environment, *The Draft NSW Planning Guidelines: Wind Farms* (2012) <www.planning.nsw.gov.au/~/media/Files/DPE/Guidelines/draft-nsw-planning-guidelines-wind-farms-a-resource-for-the-community-applicants-and-consent-2011-12.ashx>; Victorian Environment and Natural Resources Committee, *Final Report on the Inquiry into the Approvals Process for Renewable Energy Projects in Victoria* (2010) Parliament of Victoria <www.parliament.vic.gov.au/component/content/article/44-inquiry-into-the-approvals-process-for-renewable-energy-projects-in-victoria/870-report>.

130 Penelope Crossley, Submission to the Department of Planning and Environment, *Draft NSW Planning Guidelines for Wind Farms*, 3 February 2012.

131 New South Wales Department of Planning and Environment, above n 129.

132 Ibid.

133 Infigen Energy, Submission to the Renewable Energy Target Review Expert Panel, *Review of the Renewable Energy Target*, May 2014, 3.

3.3.5 *Economies of Scale of the Projects Within the Renewable Energy Sector*

One of the fundamental principles of economics is that, as the rate of production increases, the incremental cost of production should decline.[134] However, the average size of projects in the renewable sector is comparatively smaller than those that rely on conventional fossil fuels. There are a number of reasons for this, which range from the infrastructure constraints identified above, to the physical resource quality or quantity, as well as other factors such as the levels of social acceptance.[135] These problems may be overcome by co-locating a number of synergistic projects within the same area so that all projects can benefit from greater economies of scale[136] and can share some of the transaction costs such as grid connection and planning permission, which may otherwise have made the projects not economically feasible.[137] However, there are some problems that cannot be quite so easily overcome. For example, many project financiers will not provide loans to a project unless a specific size is achieved due to the cost of conducting due diligence, negotiating the contract, establishing the loan and then monitoring the ongoing risk profile and creditworthiness of the project.[138]

3.3.6 *Limited Access to Finance and Capital Market Failures*

The willingness of the capital market to lend to or invest in renewable energy and the associated cost of funding will be determined by the perceived risks of the renewable energy sector relative to other sectors and markets.[139] In the aftermath of the Global Financial Crisis, there was low forward liquidity in power markets, and uncertainty surrounding future carbon prices and subsidy levels were key risk factors facing investors.[140]

Some less mature forms of renewable energy use comparatively less proven technology and have higher initial capital costs per kW than conventional fossil fuel generation. For example, 75 per cent of the cost of energy from

[134] Cosbey, above n 50, 4.
[135] Paul Curnow, Lachlan Tait and Ilona Millar, 'Financing Renewable Energy Projects in Asia: Barriers and Solutions' (2010) 1 *Renewable Energy Law & Policy Review* 101, 103–4.
[136] Cosbey, above n 50, 4; Curnow et al., above n 135, 103–4.
[137] Ottinger et al., above n 7, 189.
[138] Hamilton, above n 82, 5; see also Union of Concerned Scientists, above n 115, 21.
[139] United Kingdom Office of Gas and Electricity Markets, 'Project Discovery: Options for Delivering Secure and Sustainable Energy Supplies' (Consultation, United Kingdom Office of Gas and Electricity Markets, 3 February 2010) 1–2.
[140] Ibid.

a wind turbine is derived from 'upfront capital costs, such as the turbine purchase cost, installation of the foundation, cost of electrical equipment and grid connection'.[141] This cost is incurred prior to the project generating any revenue. In contrast, a gas-fired power station only requires 30 per cent of the cost of the energy for upfront costs, with the rest of its costs being incurred through the operation and maintenance phases.[142] This makes it more difficult for renewable energy projects to secure financing in periods of increased risk and economic uncertainty. In a British government review, the Office of Gas and Electricity Markets stated:

> In an environment of heightened, or heightened perception of, risk the cost of raising the necessary finance could become very high if the investment is to be delivered in a timely fashion, requiring prices to rise accordingly.[143]

A further constraint on the ability of renewable energy projects to acquire finance has historically been the limited ability of financiers to identify and appropriately price the risks associated with them. While this problem has now been resolved for more commercialised renewable energy sources and technologies in mature markets, it remains a problem for less developed energy sources and countries with a nascent energy sector. When this problem occurs, renewable energy projects are often faced with higher financing costs – such as interest rates, risk premiums and insurance – than conventional generation projects.[144] In the absence of regulatory support mechanisms such as subsidies or tax relief, this makes it comparatively more expensive to build and operate a renewable energy project than a conventional fossil fuel generation project with the same generating capacity.[145] Thus, renewable energy projects using less commercialised technologies in competitive markets or renewable energy projects in nascent markets may not receive a sufficient spot price to be able to repay their loans and make a profit.[146]

3.3.7 *Principal-Agent Problems/Split Incentives*

Split incentives or principal-agent problems occur where an agent undertakes activities on behalf of the principal and the two parties have different economic incentives, leading to an inefficient outcome. This problem occurs

141 Curnow et al., above n 135, 103.
142 Ibid.
143 United Kingdom Office of Gas and Electricity Markets, above n 139, 1–2.
144 See also Union of Concerned Scientists, above n 115, 21.
145 Curnow et al., above n 135, 103.
146 Edenhofer et al., above n 54, S20.

frequently in the renewable energy sector, primarily at a residential level between landlords and tenants,[147] where one party is responsible for purchasing the necessary electrical equipment and the other party has to pay the resulting energy bills. Due to the complex installation and wiring requirements, and, for on-grid renewable energy systems, the necessity of a network connection, residential renewable energy systems cannot easily be moved once they have been installed. As a result, unless the residential customer installing the system also owns the site on which it will be located, residential renewable energy systems are unlikely to be adopted. This is because most residential landlords are reluctant to invest in residential renewable energy systems as the benefits that accrue from the deployment of such a system will accrue to the tenant in the form of lower electricity bills through increased self-consumption, and possibly some income from the sale of surplus electricity to the grid. Conversely, the tenant is also unlikely to invest in a residential renewable energy system where the payback period exceeds the term of their residential lease.

3.3.8 *Lack of Skilled Labour*

A further market barrier is the result of a shortage of skilled labour within the sector.[148] The skills shortages include areas such as environmental engineers, electricians experienced in installing and upgrading the high voltage transmission and distribution lines and specialists in high tech manufacturing. While this problem has largely been addressed in developed countries, it remains a serious impediment to the development of the sector in developing countries.[149] This problem is particularly pronounced in rural areas.[150] A lack of skilled labour not only has an impact on the initial uptake of the technology because installation works cannot be carried out, but it also affects the ongoing operation and maintenance of the renewable energy system itself. Projects that do not have access to trained and skilled labour may not be properly operated or maintained, leading to shorter working lives and impact

147 See e.g. Sherlock and Marples, above n 26; see also Brown, above n 47, 1199.

148 See generally RenewableUK, *Skills Manifesto: Skills Policy Recommendations from the Wind and Marine Energy Industries* (23 October 2013) <https://cdn.ymaws.com/www.renewableuk.com/resource/resmgr/publications/reports/Skills_Manifesto.pdf>; Department for Business, Innovation and Skills, Department of Energy and Climate Change, and Department for Environment, Food and Rural Affairs, 'Skills for a Green Economy: A report on the Evidence' (Paper, United Kingdom Parliament, 2011).

149 See generally IRENA, *Renewable Energy and Jobs: Annual Review 2018* (IRENA, 2018).

150 REEEP Africa, *Sustainable Energy Regulation and Policymaking Training Manual* (2010) <http://africa-toolkit.reeep.org/modules/Module3.pdf> 7.31.

on the ability of the project developer/customer to repay any associated loans.[151] This problem is so acute in India that the Indian Government had to launch the Surya Mitra Scheme in an attempt to train an additional 50,000 skilled photovoltaic technicians by March 2020.[152]

3.3.9 *Lack of Social/Consumer Acceptance*

In some communities, renewable energy projects do not have high levels of acceptance, which makes it difficult to obtain planning consent and other necessary approvals. This market barrier can largely be attributed to a lack of education/awareness and Nimbyism.[153] When the public are not widely educated about the unpriced negative externalities associated with conventional fossil fuel generation, and the unpriced positive externalities associated with renewable generation, they are unlikely to value these factors appropriately and thus are more likely to oppose renewable projects.[154]

A concern, particularly for wind energy projects, is Nimbyism.[155] The visual impact of wind turbines is subjective, with up to 67 per cent of survey respondents to one survey finding wind turbines visually appealing[156] and even 'elegant alternatives to coal and gas power plants'[157] that 'enhance the landscape'.[158] However for others, wind turbines are 'intrinsically ugly'[159] and 'ruin scenic vistas and ... previously pristine, unspoiled areas'.[160] A number of surveys have shown that residents living close to proposed commercial wind farms often fear

151 Ibid. See generally, IRENA, *Renewable Energy and Jobs: Annual Review 2018*, above n 149; IEA and IRENA, *Perspectives for the Energy Transition: Investment Needs for a Low-Carbon Energy System* (OECD/IEA and IRENA, 2017) 174–5.

152 Ministry of New and Renewable Energy, *Annual Report 2016–7*, (Government of India, 2017) 9 <https://mnre.gov.in/file-manager/annual-report/2016-2017/EN/pdf/1.pdf>.

153 This acronym stands for Not in my backyard and refers to a form of parochialism where people oppose developments in their neighbourhoods but are quite happy to accept the benefits of the developments when they are located elsewhere.

154 Brown, above n 47, 1202.

155 Kim Talus, 'Environment and Energy: On a Bumpy Road Towards a Clean Energy Future' in Kim Talus (ed.), *EU Energy Law and Policy: A Critical Account* (Oxford University Press, 2013) 175, 184.

156 ERM and REARK Research, *Study into Community Attitudes About Wind Farms in the NSW Southern Tablelands* (Epuron, 2007) 1.

157 S Harland Butler, 'Headwinds to Clean Energy Future: Nuisance Suits Against Wind Energy Projects in the United States' (2009) 97 *California Law Review* 1337, 1341.

158 Alexandra Wawryk, 'Solar and Wind Energy' in Richard L Ottinger and Adrian J Bradbrook (eds.), *UNEP Handbook for Drafting Laws on Energy Efficiency and Renewable Energy Resources* (UN Environment Programme/Earth Print Limited, 2007) 1, 166.

159 Ibid.

160 Ernest E Smith, 'Wind Energy: Siting Controversies and Rights in Wind' (2007) 1 *Environmental Energy Law & Policy Journal* 281, 283.

their development because they believe their installation will have negative aesthetic impacts, create excessive noise or change the character of the neighbourhood. For example, a BBC Wales survey found that the installation of commercial wind farms into a previously residential or agricultural area is often of concern to neighbours, with only 32 per cent of neighbouring landowners surveyed in the BBC Wales and University of Wales survey stating they were in favour of the Taff Ely wind farm prior to the construction.[161] However, once the Taff Ely wind farm had been constructed and successfully operated for a number of years, the levels of support increased significantly, with 74 per cent of neighbouring landowners in favour of the wind farm.[162] This suggests that while this barrier may initially hamper the development of new renewable energy projects within a geographic location, acceptance among surrounding communities is likely to increase as they become more aware of the actual rather than the perceived impacts. This will inevitably mean that it is more difficult to be the first mover into a new geographic location but also suggests that this barrier may decline in importance for subsequent projects within the location. Payment to neighbouring landowners may provide sufficient incentive to facilitate the acceptance of the initial installations in new areas and prove to be an effective strategy to address this barrier.

3.4 IS REGULATORY INTERVENTION IN THE RENEWABLE ENERGY SECTOR WARRANTED FROM AN ECONOMIC PERSPECTIVE?

From a welfare economics perspective, only the existence of market failures justifies government intervention within the renewable energy sector, though the additional presence of a wide range of market barriers does compound the need for that intervention. Differences of opinion exist about the appropriate course of action for dealing with market failures.

Some economists, such as Stigler,[163] Posner,[164] Bhagwati[165] and Gordon,[166] believe that the risk of government failure, regulatory capture and imperfect

[161] Kevin Bishop and Alison Proctor, *Love Them or Loathe Them? Public Attitudes Towards Wind Farms in Wales* (BBC Wales, 1994).

[162] Ibid.

[163] George Stigler, 'The Theory of Economic Regulation' (1971) 3 *Bell Journal of Economics and Management Science* 3.

[164] Richard A Posner, 'Theories of Economic Regulation' (1974)5 *Bell Journal of Economics and Management Science* 335.

[165] Jagdish Bhagwati, *Protectionism* (MIT Press, 1988).

[166] Richard L Gordon, 'The Case Against Government Intervention in Energy Markets' (Policy Analysis No. 628, The Cato Institute, 1 December 2008) <http://object.cato.org/sites/cato.org/files/pubs/pdf/pa-628.pdf>.

information mean that regulatory intervention by the government into the sector may lead to suboptimal economic outcomes. As a result, the costs of the market failure must be balanced against the risk of government failure.[167] Despite this, it appears that the majority of economists believe that regulatory intervention within the renewable energy sector is warranted.

3.4.1 *Pigovian Taxes*

The preferred approach among many economists for addressing externalities is to internalise the external costs of the activity not currently represented in the price paid by consumers in the form of Pigovian taxes.[168] Pigovian taxes encourage the producer to reduce the external costs associated with the negative externality to the point where the marginal cost of abatement is equal to the tax rate. Conversely, where the external costs of negative externalities are not internalised, Pigovian theory supports the provision of subsidies where positive externalities exist.[169]

3.4.2 *Coase Theorem*

An alternative approach for dealing with externalities is the Coase Theorem or Contracting Theory, which states that externalities do not warrant the imposition of taxes or subsidies, but rather a more efficient outcome will be achieved when individuals are encouraged to negotiate outcomes that eliminate externalities.[170]

In reality, neither Pigovian theory nor the Coase Theorem provides a perfect solution for addressing market failures in the renewable energy sector.[171] First, while some externalities associated with conventional fossil fuel electricity generation, such as air pollution and carbon emissions, may be taxed through Pigovian taxes such as emission trading schemes devised in accordance with 'polluter pays principles', not all externalities are easy to

[167] See e.g. Talus, above n 155, 178.

[168] See e.g. Arthur Cecil Pigou, *The Economics Of Welfare* (4th edn, Macmillan, 1932); Menanteau et al., above n 22, 800; Anthony D Owen, 'Environmental Externalities, Market Distortions and the Economics of Renewable Energy Technologies' (2004) 25 *The Energy Journal* 127, 128; Yordi, above n 60; Borenstein, 'The Private and Public Economies of Renewable Electricity Generation', above n 26, 67.

[169] Pigou, above n 168.

[170] Ronald Coase, 'The Problem of Social Cost' (1960) 3 *Journal of Law and Economics* 1.

[171] Borenstein, 'The Private and Public Economies of Renewable Electricity Generation', above n 26, 87; Gordon, above n 166, 20.

price.[172] A particular problem occurs where there are multiple externalities associated with an activity, such as occurs in the energy sector, as the Pigovian taxes should all be specifically tailored. For instance, Lehmann and Gawel argue that, contrary to some previous research, the adoption of a carbon tax will not remove the need for governments to regulate to support the renewable energy sector.[173] This is because a carbon tax will not help to address all of the externalities associated with energy generation, such as energy security issues.[174] Indeed, Pigovian taxes are generally not seen to be a solution for the externalities associated with energy security.[175] Energy security, which is commonly achieved through diversifying supply and reducing dependence on imports of fossil fuels and nuclear energy, is one of the most challenging externalities to price. The value ascribed to energy security is normally a national political decision and the circumstances behind the decision are prone to change rapidly, in line with world events. For example, recent events related to sectarian conflicts in the Middle East and the Russia/Ukraine conflict have all had a direct impact on oil and gas prices, with consequent impacts for the countries that rely on imports from these regions.[176] These problems associated with Pigovian taxes are not resolved if the Coase Theorem is used instead. The Coase Theorem relies upon the assumptions of perfect competition, no transaction costs and that both consumers and producers of externalities are willing to voluntarily negotiate agreements that lead to a socially optimal resource allocation. However, these theories do not reflect the realities of the renewable energy sector. As a result, the broad consensus seems to be that the most appropriate response to address the market failures associated with the renewable energy sector is to engage in regulatory intervention in the market.

Once a decision to pursue market intervention is made, the question then becomes what form the intervention should take. Governments have a range of options available to them when intervening in a market to rectify a market failure and thus, in order to achieve the most efficient market outcome, it is important to consider the relative costs and benefits of the alternative approaches.[177] The options include maintaining the status quo, using the

172 See generally Thomas Sundqvist, 'What Causes the Disparity of Electricity Externality Estimates?' (2004) 32 *Energy Policy* 1753.

173 See also Lehmann and Gawel, above n 50.

174 Ibid.

175 Sherlock and Marples, above n 26, 6.

176 See e.g. Phillip C Beccue et al., 'An Updated Assessment of Oil Market Disruption Risks' (2018) 115 *Energy Policy* 456.

177 Borenstein, 'The Private and Public Economies of Renewable Electricity Generation', above n 26, 86–7.

existing law (possibly coupled with better monitoring), increasing enforcement, information and education campaigns, economic instruments (such as taxes, subsidies and tradable property rights), voluntary industry standards or codes of practice, self-regulation and command and control regulation.[178] There is no single model of government intervention in the renewable energy sector that will address all of the market failures and market barriers.[179] Rather, responses need to be varied to reflect the

> country or region's specific institutions, technologies, resources, economic system, political concerns and predispositions regarding market intervention.[180]

Indeed, studies have shown that a combination of tailored regulatory strategies is most effective in encouraging the growth of renewable energy generation.[181] Furthermore, in order for regulation within the renewable energy sector to be effective, it must provide a predictable long-term framework to encourage investment. This is particularly important in liberalised markets as investors make their decisions based on 'the long-term expectations of price developments and costs'.[182]

3.5 CONCLUSION

From an economic perspective, well-functioning, competitive global and domestic markets generally provide the most suitable basis for optimal and timely investment decision-making. However, electricity is a 'mixed good', possessing characteristics of both public and private goods. It is this inherent characteristic of electricity which warrants its special regulatory treatment. As shown in this chapter, the markets within the energy sector are subject to three market failures: the presence of unpriced negative and positive externalities in the energy sector; spillovers and learning effects; and information asymmetries. These market failures are further combined with a range of market barriers, which vary by country, including subsidies for fossil fuel sources and generation, regulatory uncertainty and barriers to access to the transmission and distribution networks for renewable generation.

178 See e.g. Legislation Design and Advisory Committee, 'Legislation Guidelines: 2018 edition' (March 2018) 14–16 <http://lac.org.nz/assets/Uploads/Legislation-Guidelines-2018-edition.pdf>.

179 Edenhofer et al., above n 54, S18.

180 Jaccard, above n 3, 590.

181 REN21, 'Renewable Energy Potentials: Opportunities for the Rapid Deployment of Renewable Energy in Large Economies, Its Impacts on Sustainable Development and Appropriate Policies to Achieve It' (Renewable Energy Policy Network for the 21st Century, 2008) 17.

182 Anatole Boute, 'Improving the Climate for European Investments in the Russian Electricity Production Sector' (2008) 26 *Journal of Energy & Natural Resources Law* 267, 270–1.

From a welfare economics perspective, the presence of market failures explains the need for government intervention in national energy markets to enable renewable energy to compete with fossil fuel generation on a more level playing field. However, there is no single model of government intervention in the renewable energy sector that will address all of the market failures and market barriers. Rather, a combination of tailored regulatory strategies that address the specific market failures and market barriers within the context of the country's legal, political, economic and cultural environment has been shown to be effective in encouraging the growth of renewable energy generation. In order for this government intervention to be successful, it must provide a predictable and stable long-term framework to encourage investment. Further, the intervention needs to target the strategic deployment of renewable energy within the context of the entire energy sector to ensure that there is a coordinated approach to achieving national energy security and sustainability goals.

Despite the different models of government intervention, economic theory suggests that if governments are only engaging in regulatory intervention to support electricity generation from renewable energy sources, i.e. to address the market failures, there should be a high level of normative consensus between the national renewable energy laws. This means that the reasons why countries state that they engage in regulatory intervention should be similar across different countries' renewable energy laws. This idea will be tested in the next chapter when the legislative objectives of the different countries are analysed.

4

Why Do Countries Legislate to Accelerate the Deployment of Renewable Energy?

This is the first comprehensive study of the reasons why all of the different countries with national renewable energy laws legislate to accelerate the deployment of renewable energy. This means that, previously, it was not known whether there was a normative consensus among countries as to their rationale for legislating to support the accelerated deployment of renewable energy. Do countries only legislate to correct the market failures that affect the renewable energy sector, as economic theory would suggest, or do they legislate for a far broader range of reasons that differ between them? This chapter seeks to address this gap in the literature.

4.1 WHAT DOES IT MEAN TO LEGISLATE?

Not every country has elected to legislate to support the development and accelerated deployment of renewable energy. Indeed, on 1 August 2018, more than 40 per cent of countries (86 countries out of a total of 199 countries) did not have a primary framework piece of national renewable energy law. It is hypothesised that there are two key reasons why countries do not have renewable energy legislation. First, some countries do not have the need or desire to support renewable energy, as they have high levels of energy security and competing financial interests in supporting conventional fossil fuels. Second, some countries do not have the skills, capacity or resources to develop legislation governing renewable energy.

These hypotheses appear to be borne out by the available evidence. For example, of the fifteen members of the OPEC, only Algeria, Ecuador and Venezuela have a law promoting renewable energy.[1] These countries control 82 per cent of the world's proven crude oil supplies, and all have very high

[1] OPEC, 'Member Countries' (2018) <www.opec.org/opec_web/en/about_us/25.htm>.

levels of energy self-sufficiency.[2] Equally, countries with low gross domestic products and poor access to electricity, such as Burundi, Niger, Mozambique, Central African Republic, Somalia and Liberia, are significantly less likely to legislate than countries with higher gross domestic products and greater access to electricity. Despite this, these low-income countries may still have renewable energy policies and renewable energy targets, but they will often not be binding or legally enforceable. This is a problem if countries want to achieve their policy goals, as Omorogbe has reported that laws that are specifically tailored to support the accelerated deployment of renewable energy create an environment that stimulates the growth of the sector.[3]

At this juncture, it should be noted that the market failures and market barriers that exist and prevent the full functioning of the renewable energy sector cannot be addressed by renewable energy law or policy alone. Other options exist that may also go some way towards rectifying the existing market distortions, such as imposing a carbon tax on electricity generated from fossil fuel, removing direct and indirect subsidies from fossil fuel and nuclear energy, changes to infrastructure and planning laws, priority grid connection, and taking advantage of environmental and competition laws.[4] As stated in Chapter 3, from an economic perspective the preferred approach is to internalise the costs of externalities and the other risks associated with fossil fuels and nuclear energy with Pigovian taxes.[5] However, due to the complex interplay of externalities being addressed, which range from the risks associated with carbon emissions to energy security concerns, this is not easy to achieve. This is because while some, such as Menanteau et al., have argued that the cost of environmental externalities such as the carbon emissions or impact on air quality of conventional energy use are difficult to estimate,[6] arguably these costs are easier to price than some of the other externalities associated with conventional energy use. This view is supported by the fact that as of

2 OPEC, 'World Oil Outlook' (2018) <www.opec.org/opec_web/en/publications/340.htm>.

3 Yinka O Omorogbe, 'Promoting Sustainable Development Through the Use of Renewable Energy: The Role of Law' in Donald N Zillman et al. (eds.), *Beyond the Carbon Economy: Energy Law in Transition* (Oxford University Press, 2008) 39, 45–51.

4 For a greater discussion of the role of environmental and antitrust laws, see Lincoln L Davies, 'Reconciling Renewable Portfolio Standards and Feed-in Tariffs' (2012) 32 *Utah Environmental Law Review* 311, 317.

5 Aviel Verbruggen and Volkmar Lauber, 'Basic Concepts for Designing Renewable Electricity Support Aiming at a Full-scale Transition by 2050' (2009) 37 *Energy Policy* 5732, 5741. See also Philippe Menanteau, Dominique Finon and Marie-Laure Lamy, 'Prices Versus Quantities: Choosing Policies for Promoting the Development of Renewable Energy' (2003) 31 *Energy Policy* 799, 801.

6 Menanteau et al., above n 5, 801.

1 April 2018, forty-five countries had put a national price on carbon, either through carbon taxes or emissions trading schemes,[7] with pollution taxes targeting air pollution implemented in other jurisdictions such as the Ukraine.[8] Other costs associated with conventional energy use, such as a lack of energy security or reliance on imported fuels, are much harder to price as they are largely the result of political decisions about the relative value to be ascribed to energy independence, security of supply, diversity of supply or other regional energy security approaches.

In terms of using other existing laws, Michaels has argued that national renewable energy laws may actually lead to further market distortion and over-regulation, and that the existing environmental laws could achieve a more effective outcome with greater cost-efficiency.[9] Others have similarly argued in favour of using competition laws to address issues of market dominance or abuse that may be hindering the development of the renewable energy sector.[10] However, the broad consensus appears to be that relying solely on environmental or competition laws alone would be an insufficient means to achieve the desired end. This is because the problems being addressed through such laws are not merely environmental or the result of market dominance or abuse. Rather, as will be shown in this chapter, countries support renewable energy in an attempt to overcome a wide range of problems and to achieve a variety of objectives. Thus, while over-regulation is certainly a concern and steps need to be taken in order to prevent burdensome administration and compliance costs, there is a role for separate laws promoting renewable energy that contain regulatory support mechanisms to help countries meet their legislative objectives.

So, accepting that renewable energy laws are necessary in theory, the question then becomes why specifically do particular countries regulate the renewable energy sector? The answers may largely rest in their legislative objectives.

4.2 THE IDENTIFICATION AND ROLE OF LEGISLATIVE OBJECTIVES

Berry has stated that 'a purpose section is a provision that explicitly states the social, economic or political objective or goal that is sought to be achieved,

7 World Bank and Ecofys, 'State and Trends of Carbon Pricing 2018 (May 2018, Washington DC) 17.

8 Environmental Pollution Fee 1999 (Ukraine).

9 Robert J Michaels, 'National Renewable Portfolio Standard: Smart Policy or Misguided Gesture' (2008) 29 *Energy Law Journal* 79, 86–8.

10 Paul-Georg Gutermuth, 'Regulatory and Institutional Measures by the State to Enhance the Deployment of Renewable Energies: German Experiences' (2003) 69 *Solar Energy* 205, 209–10.

assuming that the provisions of the statute are implemented by those who are required or authorised to perform that function'.[11] In this way, the legislative objectives section provides a substantive enactment which 'delimits and illuminates the legal effect'[12] of legislation. In addition, legislative objective sections are suggestive of the legislative intent, fulfilling a symbolic purpose.[13] Hence, legislative objectives provide a signal to both industry and the market about the future direction of law and policy in the sector. Where the legislative objectives are specific and measurable, they also provide a benchmark for assessing the relative success or failure of the legislation.

The primary role of legislative objectives is to act as a guide for the statutory interpretation of ambiguous legislative provisions and resolving disputes. In both common and civil law jurisdictions, when there is any uncertainty or ambiguity in the legislation, the statutory interpreter must discern the underlying purpose of the legislation (this is sometimes also referred to as 'identifying the mischief that the legislation is seeking to address'). This is an essential step in the purposive approach to construction in the common law jurisdictions, and a feature of both the teleological and exegetic approaches in the civil law jurisdictions. Once the purpose of the legislation is identified, the courts will seek to interpret the text of the ambiguous section in light of both the purpose of the legislation and in the context of the Act as a whole.[14] If two possible constructions are identified, the construction that better promotes the purpose of the legislation will be preferred.[15] It is important to note that courts will not use the purpose of the legislation to overrule the clear and unambiguous text of an operative provision.[16]

Legislative objectives often seek to fulfil other roles, such as attempting to increase support for the legislation, or to provide a description of the features of the legislation. This has prompted criticism, including from Dickerson who stated that general purpose clauses can 'degenerate into pious incantations'.[17]

11 Duncan Berry, 'Purpose Sections: Why They Are a Good Idea for Drafters and Users' (2011) *Loophole* 49, 49.

12 Ian McLeod, *Principles of Legislative and Regulatory Drafting* (Hart Publishing, 2009) 17.

13 House Legislative Counsel, *Manual on Practice and Procedure in Committees, Proceedings and Conferences of the House of Representatives* (1992) 201.

14 *IW v. City of Perth* (1997) 191 CLR 1, 12 (Brennan CJ and McHugh J).

15 See e.g. *Acts Interpretation Act* 1901 (Cth) s 15AB.

16 *Minister for Urban Affairs and Planning v. Rosemount Estates Pty Ltd* [1996] NSWCA 91 (14 August 1996). See also Evan Bell, 'Judicial Perspectives on Statutory Interpretation' (2013) 39 *Commonwealth Law Bulletin* 245, 270.

17 Reed Dickerson quoted in David Renton, *The Preparation of Legislation: Report of a Committee Appointed by the Lord President of the Council*, Cmnd 6053 (2009), cited in Ian McLeod, *Principles of Legislative and Regulatory Drafting* (Hart Publishing, 2009) 17.

Meanwhile Berry has reported that some of the legislative counsel interviewed for his research reported that a purpose or objects section can 'end up being no more than a list of grandiose statements that would not be out of place in a party political manifesto'.[18] This need not be the case if the legislative objective sections are appropriately drafted and focused on their future use in supporting the purposive interpretation of the legislation. The inclusion of other roles in the legislative objectives is strictly described as a 'misuse' of legislative objectives, with this content being more appropriately located in secondary materials such as the minister's second reading speech, and the explanatory memoranda.[19] However, given that legislation is drafted at the behest of ministers and under their ultimate control, often requiring political negotiations to secure passage by the parliament, this is unlikely to change in the near future.

4.2.1 *Identification of the Purpose of Legislation*

There are four main ways of identifying the purpose of the primary legislation governing or promoting the accelerated deployment of renewable energy. First, and most commonly, in the case of the primary framework piece of renewable energy legislation, the legislation may have a specific legislative objectives section that summarises the purpose of the legislation. Second, the purpose may be discerned from reading the legislation as a whole. Third, identification of the mischief that the legislation is seeking to remedy may assist in identifying the purpose of the legislation. Finally, the purpose of the legislation may be found from studying extrinsic materials such as *travaux préparatoires*, second reading speeches, parliamentary debates, regulatory impact statements and/or committee reports. This is not to say that any one of these ways alone will be authoritative in identifying the purpose, with a combination of these approaches often being used, particularly when competing legislative objectives have been identified.[20]

In the context of finding the purpose of national renewable energy laws, legislative objectives sections (also sometimes called the 'statement of principle' or the 'purpose section') play an important role, having largely replaced the previous practice of using the preamble to indicate the intended purpose of the legislation.[21] Almost all of the national renewable energy laws studied in

18 Berry, above n 11, 57.

19 Paul Lanspeary, *Statutory Interpretation for Drafters* (17 August 2005) Australasian Parliamentary Counsel <www.pcc.gov.au/pccconf/2005/papers/8-Paul-Lanspeary.pdf> 18.

20 *Forsyth v. Deputy Commissioner of Taxation* (2007) 231 CLR 531.

21 Dennis Charles Pearce and Robert S Geddes, *Statutory Interpretation in Australia* (LexisNexis Butterworths, 2006) 45–6.

this research had a legislative objectives section. Indeed 80 out of the 113 countries with a primary framework piece of renewable energy legislation had an objectives section, while another eight countries, as well as the IRENA and the EU, had a preamble.

4.3 THE LEGISLATIVE OBJECTIVES IN RENEWABLE ENERGY LAW

It is the study of these legislative objectives sections that helps us to understand why countries have national renewable energy laws. There have been a number of theories espoused as to why countries seek to regulate the renewable energy sector.[22] However, due to its lack of comprehensiveness, the existing literature has sought to make generalisations based on a few select case studies.[23] As a result, while some of the theories hold true globally and are reflected in the findings of this research, some do not and thus this research demonstrates that some fallacies are being perpetuated that do not reflect the available evidence.

4.3.1 *Review of Some of the Previous Research*

Previous research by Lipp has stated that 'the motivations for and objectives of RE policy are strikingly similar across most countries'.[24] This assessment is not supported by the research in this book, as shown by the significant variation between the categories of legislative objectives adopted by different countries. These categories reflect that countries have different endowments of fossil fuel and renewable resources, different energy market structures and different environmental, socio-economic and security concerns. These differences are highlighted in Table 4.1, which summarises the legislative objectives cited by the countries whose name begins with the letter 'G' that have a renewable energy law. The extent of the variation among legislative objectives exhibited in different countries' renewable energy laws will become apparent in the next section of the chapter.

22 See e.g. Aviel Verbruggen and Volkmar Lauber, 'Assessing the Performance of Renewable Electricity Support Instruments' (2012) 45 *Energy Policy* 635; Reinhard Haas et al., 'How to Promote Renewable Energy Systems Successfully and Effectively' (2004) 32 *Energy Policy* 833; John A Mathews and Erik S Reinert, 'Renewables, Manufacturing and Green Growth: Energy Strategies Based on Capturing Increasing Returns' (2014) 61 *Futures* 13; Kelly Sims Gallagher, 'Why & How Governments Support Renewable Energy' (2013) 142 *Daedalus* 59; Judith Lipp, 'Lessons for Effective Renewable Electricity Policy from Denmark, Germany and the United Kingdom' (2007) 35 *Energy Policy* 5481.

23 See e.g. Lipp, above n 22.

24 Ibid 5483.

TABLE 4.1 *Comparison of the Legislative Objectives and Their Priorities of the Countries Whose Name Begins with the Letter 'G'*

Priority of legislative objective	The Gambia	Germany	Ghana	Greece	Guatemala
1	Energy security	Sustainable development	More efficient use of natural resources and energy conservation	Meet international treaty obligations and international agreements	Promote private investment and FDI
2	Reduce use of fossil fuels or nuclear imports	Reduce GHGs and address climate change	Environmental protection	Diversify supply	
3	Diversify supply	Environmental protection	Sustainable development	Energy security	
4	Affordable energy	Affordable energy	Promote private investment and FDI	Strengthen the economy	
5	Reduce GHGs and address climate change	Reduce use of fossil fuels or nuclear imports	Diversify supply	National development	
6	Strengthen the economy	More efficient use of natural resources and energy conservation	Energy security	Promote the development of the internal energy market and regional integration	
7	Environmental protection	Encourage technological innovations	Improved access to electricity	Reduce GHGs and address climate change	
8	Create jobs or improve skills and domestic capabilities		Create jobs or improve skills and domestic capabilities		
9	More efficient use of natural resources and energy conservation				

Verbruggen and Lauber found that the priorities for renewable energy regulation in developing countries are often those of 'poverty alleviation, improvement of health and educational conditions and adaptation to a changing climate',[25] all of which require a wider access to renewable electricity. In contrast, they argue that in 'industrialized countries, carbon dioxide emissions mitigation and security of supply are the main drivers of a growing interest in renewable electricity'.[26] This book shows that these assumptions are not all correct. The research in this book certainly supports Verbruggen and Lauber's contention that low and lower-middle income countries[27] are more likely than high income countries[28] to be concerned about access to electricity and improving living conditions. However, in terms of poverty alleviation, high-income countries are significantly more likely than low to lower-middle income countries to cite 'affordable energy' as a legislative objective. Further, while they were correct in their assessment that high income countries were more likely to prioritise reducing greenhouse gas emissions and climate change mitigation than low and lower-middle income countries, their statement about security of supply is not supported by the available evidence. In fact, as shown in detail by Table 4.3 (see page 113), low and lower-middle income countries are more likely than high-income countries to list 'energy security' as a legislative objective.

Other research by Haas et al. stated that the central purpose of encouraging renewable generation is to support the substitution of fossil fuels for renewable energy. The 'objectives derived from this core objective are: (i) to stimulate technological progress; (ii) to trigger learning effects with respect to investment costs; (iii) to minimise administration and transaction costs; (iv) to maintain public acceptance regarding RES technologies'.[29] In a similar vein, Mathews and Reinert argue that because renewable energy systems

> embody technological change, manufacturing, learning curve effects, and the capture of increasing returns ... Renewables may be viewed as a developmental strategic choice – and effects on climate change mitigation, energy security and environmental cleanliness are useful and desirable adjuncts.[30]

[25] Verbruggen and Lauber, 'Assessing the Performance of Renewable Electricity Support Instruments', above n 22, 636.

[26] Ibid.

[27] Countries are organised according to GNI per capita levels as follows: 'high' is $US12,056 or more, 'upper-middle' is $US3,896 to $US12,055, 'lower-middle' is $US996 to $US3,895, and 'low' is $US995 or less. Per capita income levels and group classifications from World Bank, 2018. The World Bank, World Bank Country and Lending Groups (2018) <https://datahelpdesk.worldbank.org/knowledgebase/articles/906519>.

[28] Ibid.

[29] Haas et al., above n 22, 834.

[30] Mathews et al., above n 22, 13.

The research in this book does not support the contentions of either Haas et al. or Mathews and Reinert that these objectives reflect the primary reasons why countries intervene in the renewable energy sector. This is because countries tend to refer to these objectives less frequently than others, such as environmental protection or energy security, and when they do mention industrial policy objectives or 'national development' as a legislative objective, they tend to be prioritised lower than other categories of objectives. For example, only five countries cite local manufacturing as a legislative objective, only twelve countries target national development, only thirteen countries target job creation, while an objective targeting 'technological innovation' is only found in about 19 per cent of countries with renewable energy laws.

Sims Gallagher's research identified four categories of motivations for the national promotion and development of renewable energy:

1. economic motives such as encouraging economic growth, job creation and manufacturing of renewable technologies;
2. a high endowment of renewable energy sources or a low endowment of fossil fuel sources;
3. political motivations which encourage renewable energy deployment; and
4. cultural values and attitudes supportive of renewable generation.[31]

She noted that different countries and states prioritise these motivations to different extents, reflecting their diverse issues.[32] While this is all correct, Sims Gallagher ignores some of the highest priorities that countries reflect in their legislative objectives. For example, some of the notable exceptions from her list include sectoral objectives such as improving system safety and reliability, environmental objectives such as seeking to reduce greenhouse gas emissions and mitigate the impacts of climate change, and security objectives such as diversifying supplies. These results are shown in the research detailed below.

The research conducted by Aguirre and Ibijunle analysed the renewable energy deployment in thirty-eight countries between 1990 and 2010 and compared it to a range of quantitative factors such as CO_2 emissions and energy imports to assess the main drivers of renewable energy growth. They concluded that 'over the past couple of decades, environmental concerns have been more critical drivers of countries' decisions to increase renewables investment than energy security'.[33] This view is also echoed by Charnovitz

31 Sims Gallagher, above n 22, 60–1.

32 Ibid.

33 Mariana Aguirre and Gbenga Ibijunle, 'Determinants of Renewable Energy Growth: A Global Sample Analysis' (2014) 69 *Energy Policy* 374, 380.

and Fischer, who stated, 'renewable energy policies enjoy broad public and governmental support around the world . . . the foremost rationale for them is to reduce the threat of climate change'.[34] This result is interesting because it is not reflected in the national renewable energy laws, which clearly prioritise energy security over environmental concerns. This may be explained by the sample chosen by Aguirre and Ibijunle, which was the EU Member States, OECD Countries and the BRIC countries, which is skewed towards wealthier countries that are more likely to be concerned about environmental protection.

It is the widespread presence of misconceptions in the existing literature, such as those found above, that highlight the need for a comprehensive study of the legislative objectives in national renewable energy laws. This will be undertaken in the next section of the chapter.

4.4 A COMPREHENSIVE STUDY OF THE LEGISLATIVE OBJECTIVES IN RENEWABLE ENERGY LAWS

As with most legislation, countries seek to achieve multiple objectives through their renewable energy law. Indeed, while the economic justifications for legislating for renewable energy are reasonably contained, in the legislation studied for this research there were twenty-eight different categories of legislative objectives identified. This represents a much broader range of legislative objectives than has been identified in previous research. This diversity reflects the different problems that countries are trying to address through their legislation, as well as their different fossil fuel and renewable resource endowments, their level of economic development, and the level of environmental awareness of their citizens.

Once the twenty-eight categories of legislative objectives were identified, these were then categorised by their primary theme. There were eight key themes identified in the legislative objectives: security, the environment, industrial policy, the economy, society, international and regional, sectoral and education and training. Many legislative objectives addressed more than one theme but for ease of analysis and discussion they have been addressed by their primary theme. For example, while it is arguable that 'sustainable development' could relate to both the economic and environmental themes, the economic theme was identified as the primary theme based on the text and context of the specific legislative objective sections addressing this issue.

34 Steve Charnovitz and Carolyn Fischer, 'Canada – Renewable Energy: Implications for WTO Law on Green and Not-so-Green Subsidies' (2015) 14 *World Trade Review* 177, 185.

Table 4.2 details the total number of countries (out of the 113 countries with a national renewable energy law) with each specific category of legislative objective contained in their legislative objectives section or preamble (as appropriate). In addition, the table also provides a weighted rank for each specific category of legislative objective. This reflects Talus's view that:

> The key determinant of a list of objectives, though, is not the rarely justifiable declaration of a lofty overall goal, but rather its rank and weight within often conflicting objectives.[35]

To calculate the weighted rank, a rank order number was assigned for each legislative objective mentioned in the legislative objectives section. For example the first category of legislative objective mentioned is assigned a 1, the second a 2, and so on. For example, s 3 of the Australian Renewable Energy (Electricity) Act 2000 states:

> The objects of this Act are:
>
> (a) to encourage the additional generation of electricity from renewable sources; and
> (b) to reduce emissions of greenhouse gases in the electricity sector; and
> (c) to ensure that renewable energy sources are ecologically sustainable.

The following ranks order numbers were assigned to this section:

- s 3(a) was assigned no number as this reflects the general purpose of all renewable energy legislation to deploy additional renewable energy;
- a '1' was assigned to the category of 'reduce greenhouse gas emissions and address climate change' on the basis of s 3(b); and
- a '2' assigned to the category of 'environmental protection' on the basis of s 3(c).

Once the rank order numbers had all been noted, all of the numbers in that category's column were then added, before being divided by the number of countries that cited that specific category of legislative objective. This produced the weighted rank number for each category of legislative objective.

35 Kim Talus, 'Environment and Energy: On a Bumpy Road Towards a Clean Energy Future' in Kim Talus (ed.), *EU Energy Law and Policy: A Critical Account* (Oxford University Press, 2013) 175, 187.

TABLE 4.2 *Legislative Objectives by the Number of Countries Adopting Them and Weighted Rank*

Theme	Category of legislative objectives	Number of countries citing the legislative objective (out of 113 countries plus the IRENA)	Weighted rank (rank order in list of objectives cited/frequency of citation)
Security objectives	Energy security	49	3.41
	Diversify supply	41	3.73
	Reduce use of fossil fuels imports or nuclear imports	35	3.69
	Encourage greater use of indigenous energy sources	10	3.20
	Subtotal citations/Average weighted rank	**135**	**3.56**
Sectoral objectives	Improve energy system safety and reliability	33	4.03
	Improve the structure of the energy sector	28	3.82
	Subtotal citations/Average weighted rank	**61**	**3.93**
Economic objectives	More efficient use of natural resources	46	4.22
	Sustainable development	37	4.08
	Competition and consumer issues	26	4.92
	Promote private investment and FDI	22	6.00
	Strengthen the economy	21	4.62
	National development	12	4.42
	Increase the number of IPPs and small and medium enterprises	9	7.78
	Subtotal citations/Average weighted rank	**173**	**4.76**
Education and research objectives	Encourage research	11	5.09
	Increase information about RES/public education	5	4.40
	Subtotal citations/Average weighted rank	**16**	**4.87**

TABLE 4.2 (*continued*)

Theme	Category of legislative objectives	Number of countries citing the legislative objective (out of 113 countries plus the IRENA)	Weighted rank (rank order in list of objectives cited/ frequency of citation)
International/regional objectives	Promote the development of the internal energy market and regional integration	17	5.18
	Meet international treaty obligations and international agreements	11	5.55
	Subtotal citations/Average weighted rank	**28**	**5.32**
Environmental objectives	Environmental protection	55	5.09
	Reduce greenhouse gas emissions and address climate change	28	5.89
	Reduce risk of natural and nuclear disasters	2	6.50
	Subtotal citations/Average weighted rank	**85**	**5.38**
Industrial policy objectives	Support the development of new industry and infrastructure	24	5.38
	Encourage technological innovations	22	4.95
	Create jobs or improve skills and domestic capabilities	13	6.23
	Local manufacturing	5	5.20
	Subtotal citations/Average weighted rank	**64**	**5.39**
Social objectives	Public health or improving living standards or social development	19	7.11
	Improved access to electricity	17	5.06
	Affordable energy	13	4.31
	Promote rural development	11	7.45
	Subtotal citations/Average weighted rank	**60**	**5.98**

The average weighted rank provides an indication of the relative weight that the countries that have adopted that category of legislative objective place on that category. This means that the closer the weighted rank is to 1, the higher the priority placed on that legislative objective by the average country that has adopted that specific category within its legislative objectives section.

This is a valuable tool as previous research by Grace et al. found that 'identifying clear objectives and signalling their proto-station or weight directly influences both the effectiveness of implementation and the degree of conflict during implementation'.[36] This is because it enables the implementing parties and stakeholders to understand and apply the legislation within the context of the priorities assigned to the legislative objectives and any constraints attached to it. Where this does not happen, research suggests that it makes monitoring and compliance more difficult, one objective may dominate to the exclusion of others and disputes may arise around the legislative interpretation.[37] Over time, this tool may be used to uncover trends about how the legislative priorities are shifting as countries amend their legislation or new countries enact legislation.

The results of this research are important. First, they highlight that more countries address the issue of environmental protection in their legislative objectives (fifty-five countries) than energy security (forty-nine countries). However, they also show that the weighted rank of environmental protection at 5.09 is significantly lower than that of energy security at 3.41, suggesting that the legislative objective of environmental protection is on average assigned a lesser priority among those countries that have adopted each objective. Indeed, given that the average country has between five and six legislative objectives and the weighted ranks of 'environmental protection' and 'reduce greenhouse gas emissions and address climate change' are 5.09 and 5.89 respectively, this suggests that many countries may simply be tacking what are arguably important objectives on the end of their legislative objectives sections as a means of capturing the diversity of political opinion in the community.

Table 4.2 also produces a total number of citations and weighted rank for each broad theme found within the legislative objectives. It was necessary to categorise the legislative objectives by theme due to the overlapping nature and often quite similar outcome of various categories. For example, while

36 Robert C Grace, Deborah A Donovan and Leah L Melnick, *When Renewable Energy Policy Objectives Conflict: A Guide for Policymakers* (2011) National Regulatory Research Institute <http://nrri.org/download/2011–17-when-renewable-energy-policy-objectives-conflict/#> 23.

37 Ibid.

countries differentiate between 'energy security', 'diversify supply' and 'reduce fossil fuel and nuclear imports', arguably these are designed to achieve a similar outcome, and so have been grouped to reflect this. The next section of this chapter will separately analyse each category of legislative objective identified through this research. The categories have been arranged by their broad theme.

4.4.1 *Security Objectives*

There were four legislative objectives identified in the renewable energy laws studied that targeted the theme of security:

1. energy security;
2. diversify supply;
3. reduce use of fossil fuel imports or nuclear imports; and
4. encourage greater use of indigenous energy sources.

The theme of security was the second most commonly identified in the legislation with 135 citations, but overall received the highest weighted rank of any theme at 3.56. The high priority placed on this theme is likely to reflect that these categories of legislative objectives are seeking to address a source of market failure, that is, the unpriced cost of ensuring a secure supply of energy.

4.4.1.1 Energy Security

Energy security has been defined by the IEA as energy that is 'adequate, affordable and reliable'.[38] The European Commission (EC) has adopted a similar but more detailed definition of energy security as 'the uninterrupted physical availability of energy products on the market at a price which is affordable for all consumers (private and industrial)'.[39] Barton et al. define it as 'a condition in which a nation and all, or most, of its citizens and businesses have access to sufficient energy resources at reasonable prices for the foreseeable future free from serious risk of major disruption of service'.[40]

38 Samantha Olz, Ralph Sims and Nicolai Kirchner, *Contribution of Renewables to Energy Security* (IEA, 2007) 13.

39 European Commission, *Green Paper: Towards a European Strategy for the Security of Energy Supply* COM(2000) 769 Final (29 November 2000). See also, European Commission, *Communication from the Commission to the Council and the European Parliament: European Energy Security Strategy* COM(2014) 330 final, Brussels (28 May 2014).

40 Barry Barton, Catherine Redgwell, Anita Rønne and Donald Zillman (eds.), *Energy Security: Managing Risk in a Dynamic Legal and Regulatory Environment* (Oxford University Press, 2004) 5.

The need to ensure a secure supply of energy in light of diminishing reserves of fossil fuels is a central concern of both developed and developing countries. The countries, in Table 4.3, all target energy security within their legislative objectives. These legislative objectives targeting energy security were prioritised highly, with an average weighted rank of 3.41.

TABLE 4.3 *Countries Citing 'Energy Security' in Their Legislative Objectives*

Rank of objective	Countries
1	Barbados, Colombia, Estonia, The Gambia, Japan, Jordan, Macedonia, Morocco, Palau, Philippines, Romania, Russia, South Africa, Sri Lanka, Syria, Ukraine, World (IRENA).
2	Armenia, China, Denmark, Dominican Republic, Kosovo, Paraguay, Tajikistan, Thailand.
3	Austria, Cameroon, Cote d'Ivoire, Finland, Greece, Indonesia, Palestine, Switzerland, Togo.
4	Bangladesh, Kyrgyzstan, Senegal.
5	France, Iceland.
6	Ghana, Latvia.
7	Madagascar.
8	South Korea.
9	Albania, The Bahamas, Bulgaria, Lithuania.
10	Peru.
12	Croatia.

A number of geopolitical and economic factors have led to an increased emphasis on energy security in recent years. For example, there have been growing concerns expressed about the vulnerability of Europe given their heavy dependence on Russian gas,[41] particularly in light of past and more recent conflicts with the Ukraine. Equally, the sectarian violence in parts of the Middle East such as Iraq and Syria, concerns about Iran's nuclear programme and events such as the Arab Spring have led to anxieties about supply disruptions.[42] Further issues have been created by the 'grab for resources' by

[41] Carolyn Fischer and Louis Preonas, 'Combining Policies for Renewable Energy: Is the Whole Less than the Sum of Its Parts?' (2010) *International Review of Environmental and Resource Economics* 51, 53.

[42] See e.g. Phillip C. Beccue et al., 'An Updated Assessment of Oil Market Disruption Risks' (2018) 115 *Energy Policy* 456; Steven Griffiths, 'A Review and Assessment of Energy Policy in the Middle East and North Africa Region' (2017) 102 *Energy Policy* 249, 251.

China and India as they seek to secure access to foreign sources of supply to meet their rising energy demand.[43]

It is important to note that for most countries, the primary basis for their concern about energy security is their poor level of energy self-sufficiency. There are a few outlying energy exporters that are either energy self-sufficient or very close to energy self-sufficient, such as Colombia (369 per cent energy self-sufficiency), Russia (188 per cent), Indonesia (189 per cent), Cameroon (137 per cent), Paraguay (131 per cent), South Africa (118 per cent), pre-war Syria (113 per cent),[44] Ghana (106 per cent), Peru (104 per cent), Estonia (102 per cent), Denmark (99 per cent), Cote d'Ivoire (97 per cent) and Albania (95 per cent).[45] However, the remaining thirty-six countries all have low levels of energy self-sufficiency. Indeed, the latter group have an average level of 50 per cent of their energy needs being met with domestic energy sources.[46]

Amongst energy exporters, these countries seek to address energy security in their legislative objectives for two reasons. First, some countries such as Denmark need to promote renewable energy because while they export crude oil (exporting 6.5 per cent of their domestic crude oil production each year),[47] 62.9 per cent of their domestic electricity comes from renewable energy.[48] These figures suggest that if the high levels of renewable generation are not maintained and the levels of oil production continue to decline, Denmark will quickly become a net importer of energy. Interestingly, the introduction of the Danish law also coincided with record oil prices in 2008, and thus it may also reflect a concern with improving their terms of trade. This reflects the second reason why energy exporters may be concerned about energy security. That is, it appears to be a common phenomenon that the energy exporters view energy security as being focused on maintaining the 'security of demand' for their energy exports due to the important contribution it provides to their government revenues.[49]

43 See e.g. Brye Butler Steeves and Helton Ricardo Ouriques, 'Energy Security: China and the United States and the Divergence in Renewable Energy' (May/August 2016) 38(2) *Contexto Internacional* 643; Gawdat Bahgat, 'Europe's Energy Security: Challenges and Opportunities' (2006) 82 *International Affairs* 961, 961.

44 2009 data. Source: IRENA, *Renewable Energy Country Profiles Middle East* (IRENA, 2012) 25.

45 Author's own calculations based on the ratio of production to total primary energy supply in 2016, underlying data derived from IEA, *World Energy Balances* (IEA, 2017).

46 Ibid.

47 IEA, *Energy Policies of IEA Countries, Denmark 2017 Review* (2017) 43.

48 Ibid 135 (2016 data).

49 Daniel Yergin, 'Ensuring Energy Security' (2006) 84 *Foreign Affairs* 69, 70–1. See also International Finance Corporation, *Renewable Energy Policy in Russia: Walking the Green Giant* (IFC, 2011).

Among energy importers, there are a number of ways that countries may address short- and long-term energy insecurity including by 'establishing strategic reserves, diversification of sources of supply and their origin, switching to renewable energy, technological innovation, and so on'.[50] Increasing the share of renewable energy is often an important strategy towards improving a country's energy security and reducing the risk of fossil fuel price volatility. For instance, these issues are addressed within the Gambian Renewable Energy Act 2013:

> Objects and reasons
>
> (a) promote the use of renewable energy resources, including hybrid systems, to achieve greater energy self-reliance and thereby reduce exposure to fossil fuel price fluctuations, reduce harmful emissions, and promote economic growth and protection of environment in the Gambia.[51]

4.4.1.2 Diversify Supply

Diversifying supply was the second most common category within the security theme, with forty-one countries identifying it as a legislative objective. It was also reasonably highly weighted, with an average weighted rank of 3.73.

TABLE 4.4 *Countries Citing 'Diversify Supply' in Their Legislative Objectives*

Rank of objective	Countries
1	Dominican Republic, Kazakhstan, Mexico, Panama, San Marino, Taiwan, Togo.
2	Cuba, Finland, Greece, Kyrgyzstan, Mauritius, Morocco, Palau, Palestine, Philippines, Switzerland, Ukraine, Yemen.
3	Afghanistan, Czech Republic, The Gambia, Paraguay, Romania, South Africa, Uzbekistan.
4	Cote d'Ivoire, Turkey.
5	Ghana, South Korea, Latvia, Senegal.
6	The Bahamas.
7	Macedonia, FYR, Serbia.
8	Albania, Honduras, Lithuania.
9	Peru, World (IRENA).
11	Croatia.

[50] Neil Gunningham, 'Confronting the Challenge of Energy Governance' (2012) 1 *Transnational Environmental Law* 119, 124.

[51] This section is unnumbered within the legislation.

The main reason why countries seek to diversify their energy mix is to reduce the risk and subsequent consequences of a disruption either in the supply of specific fuels or from specific suppliers. Many of the countries that legislate to diversify their supply are vulnerable to supply shocks due to heavy dependence on energy supplies from a single country. For example, Finland imports 100 per cent of its gas and 88 per cent of its oil from Russia,[52] the Czech Republic imports 100 per cent of its gas[53] and 66 per cent of its oil[54] from Russia, while Japan obtains 85 per cent of its oil from OPEC producers in the Middle East.[55] For these countries, increasing the supplies of renewable energy in their energy mix may act as a powerful hedge against future price fluctuations or oil shocks. Interestingly, many of the Acts targeting diversity of supply such as those from Kyrgyzstan, Peru, Romania, Taiwan and South Africa were introduced either during or in the immediate aftermath of the oil shocks in 2008.

Renewable energy does not simply remove the risks associated with importing foreign fuels or price volatility. Rather, diversifying the energy mix may lessen 'technology risks, secure additional reliability benefits, support the expansion of the renewable energy industry, spread the local benefits around geographically, advance less mature technologies so that they are less costly when other lower-cost technology potential is tapped out, or respond to expressed preferences of the public'.[56]

An example of a legislative objective targeting diversifying supply is found in the preamble to the Peruvian Legislative Decree on Investment Promotion for Electricity Generation with the Use of Renewable Energy 2008, which focuses on:

> Promoting renewable energy, removing all barriers and obstacles to its development, as a means of promoting diversification of the energy mix, which is a step towards a policy of secure energy supply and protection of the environment.

4.4.1.3 Reduce Use of Fossil Fuels Imports or Nuclear Imports

Thirty-five countries have a legislative objective designed to reduce the use of fossil fuel imports or nuclear imports (see Table 4.5). Again, this objective is weighted reasonably highly, with an average weighted rank of 3.69.

52 IEA, *Finland – Energy System Overview 2018* (IEA, 2018) <www.iea.org/media/countries/Finland.pdf>.

53 IEA, *Czech Republic – Energy System Overview 2018* (IEA, 2018) <www.iea.org/media/countries/CzechRepublic.pdf>.

54 IEA, *Oil Information 2017* (OECD/IEA, 2017) III.150.

55 Ibid III.316.

56 Grace et al., above n 36, 4.

TABLE 4.5 *Countries Citing 'Reduce Fossil Fuel Imports or Nuclear Imports' in Their Legislative Objectives*

Rank of objective	Countries
1	Cuba, Denmark, Palestine, Senegal.
2	Afghanistan, Cote d'Ivoire, The Gambia, Honduras, Mexico, Portugal, Romania.
3	Andorra, Bangladesh, Croatia, Dominican Republic, Kazakhstan, Mauritius, Panama, Philippines, Ukraine.
4	Albania, El Salvador, Japan, Uzbekistan.
5	Algeria, Germany, Kosovo, Tajikistan.
6	France, Lithuania, Serbia.
7	Austria, Latvia.
8	Macedonia, FYR, Peru.

Greater use of renewable energy sources reduces reliance on foreign imports of fossil fuels or nuclear feedstocks. This makes countries less vulnerable to price volatility and reduces the impact of external fuel shocks. In addition to these economic and energy security concerns, countries also often cite environmental and public health concerns as a reason for shifting from fossil fuel and nuclear power generation. For example, Article 1 of the Honduran Law for the Promotion of Electricity Generation with Renewable Resources 2007 states:

> This Law has the principal aim of promoting public and/or private investment in electricity generation projects with national renewable resources, through fulfilment of the following objectives: (1) to facilitate investment in and development of projects using renewable energy resources, enabling dependence on imported fuels to be reduced through the use of the country's renewable energy resources that are compatible with the conservation and improvement of natural resources.

Another example of a legislative objective seeking 'to reduce fossil fuel or nuclear imports' is contained in s 2 of the Philippines' Renewable Energy Act of 2008 (Republic of the Philippines), which states:

> It is hereby declared the policy of the State to (a) accelerate the exploration and development of renewable energy resources . . . to reduce the country's dependence on fossil fuels and thereby minimize the country's exposure to price fluctuations in the international markets, the effects of which spiral down to almost all sectors of the economy.

4.4.1.4 Encourage Greater Use of Indigenous Energy Sources

Only ten countries stated that increasing the use of indigenous energy sources was a legislative objective for them. This is likely to be because other countries have addressed this issue either through their objectives to 'ensure energy security', or 'diversify supply'. The countries that have adopted this approach do have a significantly lower GDP per capita (2016) on average of $US10,197, especially when compared to the average for all countries with renewable energy laws: $US16,024, those seeking to ensure energy security: $US13,581, those seeking to diversify their supply: $US11,557 and those seeking to reduce use of fossil fuel imports or nuclear imports: $US12,373.[57]

TABLE 4.6 *Countries Citing 'Encourage Greater Use of Indigenous Energy Sources' in Their Legislative Objectives*

Rank of objective	Countries
1	Andorra, Cabo Verde, Cote d'Ivoire.
2	Croatia.
3	Armenia, Portugal.
4	Indonesia, Philippines.
5	El Salvador.
8	Latvia.

The Andorran Law on the Promotion of Economic Activity and Social and Rationalisation and Optimisation of Resources Administration 2010 provides an example of this type of legislative objective in Article 19(1):

> with the aim of contributing to the utilisation of national natural resources and reducing dependence on external energy sources.

4.4.2 *Sectoral Objectives*

Legislation targeting the renewable energy sector is often designed to address broader sectoral issues within the energy sector that relate to the day to day functioning of the national electricity market. Objectives such as these may assist with maintaining system safety and reliability by making the process of

[57] Author's own calculations based on World Bank data on GDP per capita 2016 (current $US) rounded to the nearest whole dollar. (Source: The World Bank, *World Development Indicators* (2018) <data.worldbank.org/indicator/NY.GDP.MKTP.CD>.

balancing energy demands against intermittent supply easier. These legislative objectives may also seek to alter the structure of the energy sector through, for example, encouraging net metering or more distributed generation to reduce reliance on large fossil-fuelled power plants. Due to the high importance that countries place on having an efficient and effective energy sector, the sectoral theme of legislative objectives had the second highest weighted rank after the security objectives, with an average weighted rank of 3.93.

4.4.2.1 Improve Energy System Safety and Reliability

Energy system safety and reliability focuses on mitigating or reducing risks that may cause the power network to become overloaded or fail. This objective is designed to reduce the number of power outages, improve the technical performance of electricity generation, transmission and distribution and increase safety. Initiatives in this area assist the transmission operator in balancing the supply and demand on the energy networks, as well as in ensuring that a stable voltage is maintained.

This legislative objective was adopted by thirty-three countries (Table 4.7) and was often highly prioritised, with an average weighted rank of 4.03:

TABLE 4.7 *Countries Citing 'Improve Energy System Safety and Reliability' in Their Legislative Objectives*

Rank of objective	Countries
1	The Bahamas, Kosovo, Montenegro, Serbia, Slovenia, Switzerland, Thailand, Yemen.
2	Barbados, Ecuador, Latvia, Macedonia, FYR, The Netherlands, Senegal, South Africa, Turkey.
3	Japan, Morocco, Russia, Sri Lanka.
4	Armenia, China.
5	Hungary, Tonga.
6	Indonesia, Jamaica, Madagascar.
7	Paraguay.
9	Kazakhstan.
10	Bulgaria, Lithuania.
12	Peru.
13	El Salvador.

When this objective is prioritised, large quantities of distributed renewable energy generation can increase system safety and reliability by decreasing the load on the network. In addition, more dispersed sources of generation can make the electricity network less vulnerable to widespread blackouts.[58] The introduction of large quantities of variable or intermittent renewable generation does, however, pose its own challenges in terms of system balancing and voltage management. This makes it a complex but important task, as improving energy system safety, quality and reliability plays an important role in the industrial development of emerging economies.

An example of this legislative objective is found in the Yemeni law:

> The purpose of this law shall be to achieve the following: [. . .] E-Ensure the safety, continuity and quality of the electricity service.[59]

4.4.2.2 Improve the Structure of the Energy Sector

Negro et al. have defined market structure as 'the organisation of the current market and the criteria used to select innovation'.[60] In most countries, the market structure of the energy sector favours the incumbent and dominant market players, who, in this case, are fossil fuel generators. This acts as a barrier to entry to the introduction of renewable energy sources and new market participants to the energy market.[61]

Twenty-seven countries have identified improving the structure of the energy sector as an objective of their renewable energy laws (see Table 4.8), with an average weighted rank of 3.82. These countries tend to be low to upper-middle income countries.

This objective is closely linked to the objectives targeting competition, increasing the number of independent power producers and small to medium-sized enterprises engaged in generating electricity, improving system safety and reliability and encouraging the development of new industries and infrastructure.

While this objective is vague, there are a number of ways that countries are currently attempting to improve the structure of their energy sectors. One

58 United States Environmental Protection Agency, *Assessing the Multiple Benefits of Clean Energy – A Resource for States* (US EPA, 2011) 55.

59 «قانون رقم (1) لسنة 2009م بشأن الكهرباء» [Electricity Law 2009] (Yemen) ch 3 Art 4 [Linguistico Translations translation from Arabic].

60 Simona A Negro, Floortje Alkemade and Marko P Hekkert, 'Why Does Renewable Energy Diffuse So Slowly? A Review of Innovation System Problems' (2012) 16 *Renewable and Sustainable Energy Reviews* 3836, 3838.

61 Dörte Fouquet, 'Policy Instruments Renewable Energy – From a European Perspective' (2013) 49 *Renewable Energy* 15, 17.

TABLE 4.8 *Countries Citing 'Improve the Structure of the Energy Sector' in Their Legislative Objectives*

Rank of objective	Countries
1	Cameroon, Jamaica, Kyrgyzstan, Latvia, Madagascar.
2	Japan, Lichtenstein, San Marino.
3	China, Democratic Republic of the Congo, Pakistan, Peru, Serbia, Syria.
4	Ecuador, Macedonia, FYR, Morocco, Taiwan, Yemen.
5	Cote d'Ivoire, Russia, Thailand.
6	Netherlands.
7	Cuba.
8	Kazakhstan.
9	Honduras.
10	The Bahamas.

example of this is the move towards distributed renewable generation. This uses smaller units for electricity generation and places them closer to the end user, thereby reducing reliance on large power plants and avoiding load losses as the electricity does not need to be transmitted over long distances. In this way, the costs of fossil fuel feedstocks are avoided (which also provides a hedge against fossil fuel price rises), upgrades to large-scale power plants may be deferred or avoided altogether and the costs of upgrading or building new transmission lines may be deferred or avoided.

One downside to using more distributed generation is that, at present, distributed electricity is more costly to produce than that produced using large power plants. There may also be implications for system balancing if large amounts of small-scale distributed generation are added into the energy mix, as there may still need to be back-up generation capacity installed for planning and operating reserves, i.e. the times that the distributed generation is not supplying a sufficient quantity of electricity to meet the consumer demand. This is most likely to occur during extreme weather events when electricity demand peaks. A further concern may also relate to the implications that increasing the proportion of distributed generation may have for system safety and reliability.

An example of this objective is found in the Renewable Energy Law of the People's Republic of China 2005:

> This Law is enacted for the purpose of promoting the development and utilisation of renewable energy, increasing the supply of energy, improving

the structure of energy, safeguarding the safety of energy, protecting environment and realising a sustainable economic and social development.[62]

4.4.3 *Economic Objectives*

During the Global Financial Crisis, the renewable energy sector was targeted as a means of stimulating the economy through the development of new technologies, new jobs, new industries and new infrastructure. Indeed, by 2010, more than $US190 billon had been pledged in the form of renewable energy support under 'Green Fiscal Stimulus' packages, predominantly by China, the United States and South Korea.[63] Countries viewed the renewable energy sector as a strategic growth industry for their industrial policies. This remains the case today, with China, Japan, the EU, the United States, India and Brazil just a few of the countries that are aggressively targeting this sector as a means of achieving their economic and other goals.

The economic theme of legislative objectives was the third most prioritised theme in the renewable energy laws of countries, with an average weighted rank of 4.76.

4.4.3.1 More Efficient Use of Natural Resources and Energy Conservation

The legislative objective of 'more efficient use of natural resources and energy conservation' featured in the laws of forty-six countries (as shown in Table 4.9). It was also relatively highly prioritised with an average weighted rank of 4.22.

This is likely to reflect the considerable work undertaken by a number of international organisations such as the IEA, UN-Energy, the United Nations Environment Programme (UNEP), the United Nations Development Programme (UNDP), the EU and the International Partnership for Energy Efficiency Cooperation in improving energy efficiency and energy conservation.

In the context of these renewable energy laws, energy efficiency refers to the same standard of service and performance being achieved with less energy. In contrast, energy conservation refers to the standard of service and performance being reduced due to less energy being available. As a result, energy conservation is often viewed as being faster and less costly to introduce.[64] However, due

62 中华人民共和国可再生能源法 [Renewable Energy Law of the People's Republic of China] (People's Republic of China) National People's Congress, 28 February 2005, ch 1 Art 1 [Ministry of Commerce of the People's Republic of China translation from Mandarin].

63 Bloomberg New Energy Finance, *Weathering the Storm: Public Funding for Low Carbon Energy in the Post Financial Crisis Era* (2010), 8.

64 Larry Hughes, 'The Four "Rs" of Energy Security' (2009) 37 *Energy Policy* 2459, 2460.

TABLE 4.9 *Countries Citing 'More Efficient Use of Natural Resources and Energy Conservation' in Their Legislative Objectives*

Rank of objective	Countries
1	Albania, Ghana, Lichtenstein, Tajikistan, Venezuela.
2	Andorra, Belarus, Luxembourg, Montenegro, Sri Lanka.
3	Cuba, El Salvador, Honduras, Hungary, Jamaica, Macedonia, San Marino, Slovenia, Turkey.
4	Afghanistan, Algeria, Croatia, Czech Republic, Kosovo, Netherlands, South Africa, Ukraine.
5	Austria, Bangladesh, Colombia, Estonia, Indonesia, Jordan, Serbia.
6	Armenia, Ecuador, Germany, Switzerland.
7	The Bahamas, Cote d'Ivoire, Lithuania, Thailand, Uzbekistan.
8	Philippines.
9	The Gambia, Tonga.

to it being associated with lower levels of service and lifestyle changes, energy conservation measures often do not succeed in the long-term unless education, infrastructure and pricing strategies are introduced to complement it.[65] As a result, Hughes has argued that energy efficiency measures are often more effective in reducing energy consumption over time.[66]

Examples of legislative objectives targeting 'more efficient use of natural resources and energy efficiency' include:

- 'promote economical and efficient use of energy and resources in the energy industry operation, with due consideration of the environmental impact and balance of natural resources'.[67]
- 'provide for the development, management and utilisation of renewable energy sources for the production of heat and power in an efficient and environmentally sustainable manner'.[68]
- 'identify, promote, facilitate, implement and manage energy efficient improvement and energy conservation'.[69]

65 Ibid.

66 Ibid.

67 พระราชบัญญัติ การประกอบกิจการพลังงาน พ.ศ. ๒๕๕๐ *Energy Industry Act, B E* 2550 (Kingdom of Thailand) 10 December 2007, div 1 s 7(7) [Thai Law Forum translation from Thai].

68 *Renewable Energy Act* 2011, s 1(1) (Ghana).

69 *Sustainable Energy Authority Act* 2007 (Sri Lanka) No. 35 of 2007, 18 September 2007, Preamble.

- 'State regulation in the area of support of renewable energy sources utilisation is performed with a purpose of creation of favourable conditions for electrical and (or) heat power generation with utilisation of renewable sources of energy in order to reduce energy intensity of economic sector, to decrease environmental impact of power industry, and to increase the share renewable energy sources for electrical and (or) heat power generation.'[70]

4.4.3.2 Sustainable Development

The most commonly adopted definition of sustainable development is that of the Brundtland Report: '[d]evelopment that meets the needs of the present without compromising the ability of future generations to meet their own needs'.[71] Legislative objectives targeting sustainable development were adopted by thirty-seven countries (Table 4.10), and were often quite highly prioritised, with an average weighted rank of 4.08.

TABLE 4.10 *Countries Citing 'Sustainable Development' in Their Legislative Objectives*

Rank of objective	Countries
1	Belarus, Germany, Indonesia, Lithuania, Mauritius, World (IRENA).
2	Albania, Bulgaria, Nicaragua, Serbia.
3	Algeria, Ghana, South Korea, Mexico, Montenegro, The Netherlands, Suriname.
4	The Bahamas, Colombia, France, Jamaica, Jordan, Palestine, Paraguay, Romania.
5	Czech Republic, Philippines.
6	China, Kyrgyzstan, Taiwan.
7	Estonia, Morocco.
8	Ecuador, El Salvador, Madagascar, Tonga.
10	South Africa.

70 О поддержке использования возобновляемых источников энергии Закон Республики Казахстан от 4 июля 2009 года № 165-IV [Law of the Republic of Kazakhstan No. 165-IV About Support of Use of Renewable Energy of 4 July 2009] (Kazakhstan) ch 2 Art 3(1) [Government of the Republic of Kazakhstan translation from Kazakh].

71 World Commission on Environment and Development, *Our Common Future (The Brundtland Report)* (Oxford University Press, 1987) 43.

The countries that have adopted 'sustainable development' as a legislative objective have an average GDP per capita of $US10,344, which is much lower than the average for all countries with renewable energy laws of $US16,024.[72]

One of the criticisms of the use of 'sustainable development' is that the term is often either not defined in the laws at all or defined very vaguely. This is likely to be because 'there remains no consensus of the exact meaning of the term'.[73] This may make it harder to achieve this objective than other objectives as it is often not clear what is sought to be achieved and nor is it clear what indicators will be used to measure relative success or failure. Despite this, energy plays a critical role in ensuring sustainable development in developing countries. Kaygusuz has argued that energy contributes to sustainable development in three key ways.[74] First, renewable energy has the ability to improve people's lives by placing less reliance on the collection of firewood and traditional biomass. Kaygusuz argues that this has the potential to reduce hunger, provide greater gender equality, and thereby opportunities for education, and improve health and sanitary conditions. These factors in turn are said to enhance human capital and increase participation in public governance. Second, greater use of renewable energy can support the development of critical transport and communications, and encourage the adoption of a wider range of productive economic activities, providing increased opportunities for employment for the population. Third, improving the energy sector enables greater exchanges of information, and can enhance the stability of the local business environment, as well as the national economy and society. Critically, all of these impacts are interlinked and compound to reinforce each other, meaning that the adoption of modern forms of energy such as renewable energy sources in supporting sustainable development within developing countries cannot be underestimated.

Some examples of legislative objectives focusing on 'sustainable development' include:

- 'The purpose of this law is to a) assist in the economic use of natural resources and sustainable development in the Republic of Albania.'[75]

72 Author's own calculations based on World Bank data on GDP per capita 2016 (current $US) rounded to the nearest whole dollar. (Source: The World Bank, *World Development Indicators*, above n 57.)

73 Andrea Ross, 'Why Legislate for Sustainable Development? An Examination of Sustainable Development, Provisions in UK and Scottish Statutes' (2008) 20 *Journal of Environment Law* 35, 39.

74 Kamil Kaygusuz, 'Energy for Sustainable Development: A Case of Developing Countries' (2012) 16 *Renewable and Sustainable Energy Reviews* 1116, 1119.

75 Për Burimet E Energjisë Së Rinovueshme [Law on Renewable Energy Sources] (Albania) No. 138/2013, 2 May 2013, Art 1 [Linguistico Translations translation from Albanian].

- 'The purpose of this Act is to contribute to . . . the sound and sustainable development of the national economy.'[76]
- 'The primary objectives of this Act are ... creating conditions for achieving sustainable development at regional and local levels.'[77]
- 'The purpose of this Act is to enable the energy supply to develop in a sustainable manner in particular in the interest of mitigating climate change and protecting the environment, to reduce the costs of the energy supply to the economy not least by including the long-term external effects, to conserve fossil fuel resources and to promote the further development of technologies to generate electricity from renewable energy sources.'[78]

4.4.3.3 Competition and Consumer Issues

Historically, the electricity sector was structured around vertically integrated and state-owned monopolies that were regulated through the use of 'command and control' style regulation. More recently, there has been a clear shift towards investor-owned private enterprises operating in competitive markets. This in turn has prompted concerns about the role of national governments in ensuring consumer rights and protection.

Competition and consumer issues are reflected in the legislative objectives of twenty-six countries with renewable energy laws (see Table 4.11). Half of these countries are either EU Member States, or countries officially designated as 'EU candidate countries' or, in the case of Kosovo, a 'potential candidate' country. This reflects the importance placed by the EU on competition within the energy sector for the development of the internal energy union and its inclusion in the Economic Accession Criteria.[79] This category of legislative objective has an average weighted rank of 4.92.

[76] 신에너지 및 재생에너지 개발 · 이용 · 보급 촉진법 [Act on the Promotion of the Development, Use and Diffusion of New and Renewable Energy] (Republic of Korea) No. 14670, 21 March 2017, Art 1 [Korean Legislative Research Institute translation from Korean].

[77] Закон за енергията от възобновяеми източници [Energy from Renewable Sources Act] (Republic of Bulgaria) State Gazette No. 35/3.05.2011, 3 May 2011, Art 2 [Bulgarian Government translation from Bulgarian].

[78] *Gesetz für den Ausbau erneuerbarer Energien* [Renewable Energy Source Act] (Germany) 1 January 2017, s 1(1) [German Government translation from German].

[79] EU Economic and Financial Affairs, *Economic accession criteria* (European Commission, 12 September 2018) <https://ec.europa.eu/info/business-economy-euro/economic-and-fiscal-policy-coordination/international-economic-relations/enlargement-and-neighbouring-countries/enlargement/economic-accession-criteria_en>.

TABLE 4.11 *Countries Citing 'Competition and Consumer Issues' in Their Legislative Objectives*

Rank of objective	Countries
1	Finland.
2	France, Hungary, Slovenia.
3	Bulgaria, Colombia, Ecuador, Kosovo, Thailand.
4	Estonia, Iceland, Latvia, Serbia.
5	Macedonia, FYR, Montenegro, Yemen.
6	Honduras, Morocco, Pakistan, Russia.
7	Democratic Republic of the Congo, Czech Republic, Dominican Republic.
9	Cote d'Ivoire.
10	El Salvador.
11	Lithuania.

Examples of objectives addressing competition and consumer issues include:

- 'to enhance the economy's competitive capacity through an efficient competitive electricity market'.[80]
- 'to provide ... 2) an efficient, competitive and financially sustainable energy sector, based on the principles of non-discrimination, objectivity and transparency ... 6) fulfilment of the obligations for providing a public service on the energy markets, as well as effective protection of the rights and interests of the users of energy systems, protecting the rights of the consumers of energy and especially of vulnerable consumers'.[81]

4.4.3.4 Promote Private Investment and FDI

The promotion of private investment and foreign direct investment (FDI) in the renewable energy sector is designed to reduce the dependence of the sector on government support and to help break-up the market power of the dominant market players (often former state-owned monopolies). Attracting private investment and FDI is also sometimes seen as a means of improving the standards and reliability in the sector, as foreign private investors may bring with them new, more efficient, and safer technologies. The twenty-two countries that have adopted this objective have an average GDP per capita of $US5,730. This is less than a third of the average GDP per capita for countries

80 *2007. évi LXXXVI. törvény a villamos energiáról* [Act No. LXXXVI of 2007 on Electric Energy] (Hungary) ch 1 s 1(a) [Linguistico Translations translation from Hungarian].

81 *Закон За Енергетика* [Energy Law] (Macedonia) No. 96/2018, 21 May 2018, Art 2 [Government of the Republic of Macedonia translation from Macedonian].

with renewable energy laws but without the objective ($US18,364).[82] Japan is the only high-income country with this objective. This likely reflects the fact that prior to the Fukushima nuclear disaster, Japan did not permit any foreign direct investment into its electricity sector. This legislative objective is more likely to be assigned a low priority by countries in their laws (Table 4.12), with an average weighted rank of 6.00.

TABLE 4.12 *Countries Citing 'Promote Private Investment' in Their Legislative Objectives*

Rank of objective	Countries
1	Guatemala.
2	Jordan, Madagascar, Syria.
4	Democratic Republic of the Congo, Cuba, Dominican Republic, Ghana.
5	Jamaica, Peru, Uzbekistan.
6	Yemen.
7	Kazakhstan, Tonga.
8	Cote d'Ivoire.
9	El Salvador, Japan, Montenegro, Romania, Serbia, Ukraine.
12	Morocco.

An example of such an objective is found in the Ukrainian law:

> The fundamental principles of state policy in the area of alternative energy sources are: [. . .] Attraction of domestic and foreign investment and support for entrepreneurial initiatives in the field of alternative energy sources, including through the development and implementation of nationwide and local programs for the development of alternative energy.[83]

4.4.3.5 Strengthen the Economy

There is a link between the accelerated deployment of renewable energy and GDP growth. The IRENA has reported that previous studies have found that policy interventions that encouraged renewable energy deployment led to

[82] Author's own calculations based on World Bank data on GDP per capita 2016 (current $US) rounded to the nearest whole dollar. (Source: The World Bank, *World Development Indicators*, above n 57.)

[83] Закон України Про альтернативні джерела енергії, *Law on Alternative Energy Sources* (Ukraine) 20 February 2003, No. 555-IV, Art 1 [Linguistico Translations translation from Ukrainian].

between a 0.2 and 4.0 per cent projected positive impact on GDP.[84] The IRENA has estimated that this has broad ranging impacts on the global economy, in particular if there is a comprehensive long-term energy transition 'the cumulative gain through increased GDP from 2018 until 2050 would amount to $US 52 trillion'.[85] Other areas that could signal a strengthening of the economy include lower rates for electricity, greater levels of access to electricity, job creation, new infrastructure development and support for new industries and local manufacturing. Most of these areas correlate positively in the long-term with increasing levels of renewable energy deployment (with the possible exception of job creation, as explained below). As a result, twenty countries targeted economic growth through their renewable energy laws (see Table 4.13). The average weighted rank of the legislative objectives targeting 'strengthen the economy' is 4.62.

TABLE 4.13 *Countries Citing 'Strengthen the Economy' in Their Legislative Objectives*

Rank of objective	Countries
1	Armenia, France, Hungary, Iceland, Turkey.
2	Colombia, South Korea.
3	Kyrgyzstan.
4	Bulgaria, Greece, Portugal.
5	Afghanistan, World (IRENA).
6	The Gambia, Japan, Philippines.
7	Indonesia.
8	Montenegro, Thailand.
10	Democratic Republic of the Congo.
12	El Salvador.

One of the downsides to renewable energy laws that prioritise economic objectives highly is that they may focus on renewable energy development only to the extent that it is economically advantageous in the short-term. This means that renewable energy sources and technologies that are currently least-

[84] IRENA, *Renewable Energy Benefits: Measuring the Economics* (IRENA, 2016) 15–16 <www.irena.org/documentdownloads/publications/irena_measuring-the-economics_2016.pdf>.

[85] IRENA, *Global Energy Transformation: A Roadmap to 2050* (IRENA, 2018) 47.

cost are likely to be prioritised, even if they will not be the most effective or efficient in the long-term.

Examples of this objective are found in the Armenian and Icelandic laws:

- 'The purposes of the present Law shall be defining the principles of the state policy on the development of the energy saving and renewable energy and the mechanisms of the enforcement of those aimed at: Strengthening the economic and energy independence of the Republic of Armenia.'[86]
- 'The purpose of this legislative Act is to promote an economic electricity system and thereby strengthen the Icelandic industries as well as regional development in Iceland.'[87]

4.4.3.6 National Development

Renewable energy legislation is often designed to foster national development. This is a very broad legislative objective, and it is not often clear what countries are trying to achieve by listing it in their legislative objectives, nor how any relative success or failure to meet this objective will be measured. This objective is often used in connection with some of the industrial policy objectives, such as strengthening the economy, creating jobs, local manufacturing and supporting the development of new industry and infrastructure. This approach seems to have found favour with the countries involved in high-tech manufacturing such as Japan, South Korea and Taiwan, and those seeking to increase their activity in the sector, such as Indonesia. The reference to 'national development' sometimes also appears to refer to post-conflict nation building such as in the Democratic Republic of the Congo's legislation. Indeed, the Preamble to the law from the DRC the notes that 'electricity is one of the major and irreversible factors that condition the economic, social, technological and cultural development of all nations ... '. This legislative objective had an average weighted rank of 4.42 (see Table 4.14).

86 *The Law of the Republic of Armenia on Energy Saving and Renewable Energy* (Armenia) 9 November 2004, Ch 1 Art 1 [National Assembly of the Republic of Armenia translation from Armenian].

87 *Raforkulaga* [Electricity Act] (Iceland) No. 65/2003, 15 March 2003, Ch 1 Art 1 [Government of Iceland translation from Icelandic].

TABLE 4.14 *Countries Citing 'National Development' in Their Legislative Objectives*

Rank of objective	Countries
1	El Salvador.
2	Iceland, Indonesia.
4	South Korea, Peru, Tonga.
5	Greece, Taiwan.
6	Algeria, Democratic Republic of the Congo.
7	Japan, Philippines.

Two examples of this objective are found in the Indonesian law and Japanese law, respectively:

- 'in order to support sustainable national development and improve national energy security, the management of energy shall be aimed to . . . '[88]
- 'It aims to promote the utilization [of renewable electricity], and to contribute to the strengthening of the international competitiveness of Japan, the promotion of Japanese industries, the revitalization of the region and the sound development of the national economy.'[89]

4.4.3.7 Increase the Number of IPPS and Small and Medium Enterprises

This objective is linked to other objectives targeting a more competitive electricity market and improving the structure of the electricity sector. In seeking to increase the number of independent power producers (IPPs) and the number of small and medium enterprises active within the electricity generation sub-sector, this objective is trying to reduce the market power held by the dominant players (see Table 4.15). This forms an essential component of the breakup of the traditional vertically integrated monopolies and the transition to a privatised and deregulated energy market. It also forms part of the Economic Accession Criteria to the EU.[90]

88 *Undang-Undang Republik Indonesia Nomor 30 Tahun 2007 Tentang Energi* [Law of the Republic of Indonesia Number 30 of 2007 About Energy] (Republic of Indonesia) ch 2 Art 3 [Ellen Marie O'Brien translation from Indonesian].

89 電気事業者による再生可能エネルギー電気の調達に関する特別措置法 [Act on Special Measures Concerning Procurement of Electricity from Renewable Energy Sources by Electric Utilities (Japan)] No. 108 of 2011, 30 August 2011, Ch 1 [Ministry of Justice translation from Japanese].

90 EU Economic and Financial Affairs, above n 79.

TABLE 4.15 *Countries Citing 'Increase the Number of IPPs and Small and Medium Enterprises' in Their Legislative Objectives*

Rank of objective	Countries
6	Montenegro, Thailand, Uzbekistan.
7	Russia.
8	Bulgaria, Croatia, Hungary.
10	Cote d'Ivoire.
11	Albania.

This objective is not a top priority for any country, with an average weighted rank of 7.78. An example of an objective to increase the number of IPPs and small and medium enterprises is found in the Albanian law:

> The purpose of this law is to: (f) increase the number of independent energy producers and promote the development of small and medium enterprises . . .[91]

4.4.4 *Education, Training and Research Objectives*

The theme of education, training and research objectives was only found in the legislation of sixteen countries. The average weighted rank of this theme is 4.87. Information asymmetries and positive spillovers are other identified sources of market failure that adversely affect the renewable energy sector. However, the low frequency of citation of this theme again suggests that the majority of countries are not necessarily concerned with addressing market failures through their legislation but rather prioritise other domestic concerns.

4.4.4.1 **Encourage Research**

There is a clear need for governments to support research and development within the renewable energy sector. This reflects one of the areas of market failure identified in the previous chapter. Despite this, investment in research and development into renewable energy technologies has fallen in real terms over the past seven years, with the levels of investment stagnating and inflation over the period in OECD countries being more than 10.5 per cent.[92] One of the challenges for private firms investing in research and development is that it is difficult for

[91] *Për Burimet E Energjisë Së Rinovueshme* [Law on Renewable Energy Sources] (Albania) No. 138/2013, ch 1 Art 1 [Linguistico Translations translation from Albanian].

[92] IRENA, *Renewable Energy: A Key Climate Solution* (IRENA, 2017) 7; OECD (2018), Inflation (CPI) (indicator) doi: 10.1787/eee82e6e-en (accessed 20 September 2018). See also World Economic Forum, *White Paper: Accelerating Sustainable Energy Innovation* (World Economic Forum, 2018) 6.

them to prevent the knowledge of their discoveries from either leaking or being reverse engineered by their competitors. This may affect their ability to achieve the full benefits of their investment, and hence their willingness to invest. These so-called 'knowledge spillovers' are considered in economic terms to be a positive externality. This is because where a knowledge spillover occurs, the gain to the public is greater than its private value.[93]

Eleven countries adopted a legislative objective of encouraging research into more efficient and effective renewable energy sources and technologies. On average the countries with this objective spend considerably less of their GDP on research and development (0.28 per cent) than those countries with a national renewable energy law but without the objective (1.32 per cent). Moreover, these countries also have a significantly lower average GDP per capita of $US5,394 as compared to $US17,108 for all countries with a national renewable energy law but without this objective.[94] This means that they are often dependent on external sources of research funding such as foreign aid or private research funding bodies (see Table 4.16).

TABLE 4.16 *Countries Citing 'Encourage Research' in Their Legislative Objectives*

Rank of objective	Countries
1	Afghanistan, Argentina, Tonga
3	Uzbekistan.
5	Cuba, Kazakhstan, Mauritius, South Africa.
7	Ukraine.
11	El Salvador.
13	Peru.

Nemet's research into the factors influencing cost reductions in photovoltaic solar cells indicated that technological efficiency triggered by research had more of an impact in reducing the cost of photovoltaic solar cells than subsidised market penetration.[95] This suggests that 'funding R&D in order to trigger significant technology improvements would have been a more promising avenue to efficiently achieve substantial cost reductions in early technology stages than the heavy

93 Cédric Philibert, '*Interactions of Policies for Renewable Energy and Climate*' (Working Paper, OECD/IEA, 2011) 11.

94 Author's own calculations based on World Bank data on research and development spending as a proportion of GDP and GDP per capita 2016 (current $US) rounded to the nearest whole dollar. (Source: The World Bank, *World Development Indicators*, above n 57.)

95 Gregory F Nemet, 'Beyond the Learning Curve: Factors Influencing Cost Reductions in Photovoltaics' (2005) 34 *Energy Policy* 3218, 3227–8.

subsidisation of market penetration, a policy alternative where technological improvements are rather by-products'.[96] Thus, research into renewable energy technologies plays a critical role in meeting the long-term challenges of ensuring energy security, addressing climate change and sustainably meeting rising energy demands.[97]

An example of this legislative objective is found in the Tongan law:

> The principal objects of this Act are – (a) to promote the development of the renewable energy industry in the Kingdom by: (i) researching and developing opportunities of renewable energy in the Kingdom . . .[98]

4.4.4.2 Increase Information About Renewable Energy/Public Education

Kandpal et al. have stated that 'the broad objectives of renewable energy education pertain to providing functional knowledge and understanding of facts, concepts, principles and technologies for harnessing of renewable sources of energy'.[99] Five countries sought to adopt this legislative objective, with an average weighted rank of 4.40 (see Table 4.17).

TABLE 4.17 *Countries Citing 'Increase Information About Renewable Energy/ Public Education' in Their Legislative Objectives*

Rank of objective	Countries
1	Bulgaria.
3	Lichtenstein.
4	Mauritius.
5	Belarus.
9	Ghana.

Countries that explicitly tailor their renewable energy legislation to educate the public about renewable energy sources and thereby increase public support for renewable energy may lead to different legislative outcomes than those that do not, especially if they then make decisions based on the levels of public support achieved. This may lead to renewable energy projects, which may be more cost-efficient and effective than many

96 Manuel Frondel et al., 'Economic Impacts from the Promotion of Renewable Energy Technologies: The German Experience' (2010) 38 *Energy Policy* 4048, 4056.

97 Kaygusuz, 'Energy for Sustainable Development', above n 74, 1120.

98 *Renewable Energy Act 2008* (Tonga), Art 3.

99 Tara C Kandpal and Lars Broman, 'Renewable Energy Education: A Global Status Review' (2014) 34 *Renewable and Sustainable Energy Reviews* 300, 302.

alternatives, such as onshore wind, being spurned because of a perceived lack of public support in some areas. Conversely, greater levels of support than may otherwise be warranted may be granted to small-scale domestic projects such as photovoltaic solar so that residential households and small businesses can benefit from the subsidies. Despite this, supporting popular policies may be advantageous in that they may provide the groundwork for more stringent and unpopular policies in the future.[100] They also provide a basis on which people can engage with renewable energy technologies and on which acceptance levels can be built for other renewable energy sources and technologies.

An example of this objective is found in the Ghanaian law:

> The object of this Act is to provide for the development, management and utilisation of renewable energy sources for the production of heat and power in an efficient and environmentally sustainable manner. [...] the object shall encompass ... (F) public education on renewable energy production and utilisation.[101]

4.4.5 *International/Regional Objectives*

It is surprising that only twenty-three countries have legislative objectives addressing international agreements or regional integration, especially when the number of international agreements concerning the renewable energy sector is considered. For example, there is the Statute of the IRENA, the Energy Charter Treaty, the United Nations Framework Convention on Climate Change (UNFCCC), the Kyoto Protocol, Copenhagen Accord, the Doha Amendment and the Paris Agreement, not to mention numerous other regional and bilateral agreements. This research highlights that promotion of the development of the internal energy market and regional integration is both cited more frequently, and weighted more highly, than meeting international obligations. This reflects the powerful influence of the European Union on the development of national renewable energy laws. The average weighted rank for this theme is 5.32.

4.4.5.1 Promote the Development of the Internal Energy Market and Regional Integration

This category of legislative objective was adopted by seventeen countries, with an average weighted rank of 5.18. With the exception of Paraguay and Morocco, all the countries that adopted this approach are either Member States of the EU or candidate or potential candidate countries (see Table 4.18).

[100] Rolf Wüstenhagen, Maasrten Wolsink and Mary Jean Bürer, 'Social Acceptance of Renewable Energy Innovation: An Introduction to the Concept' (2007) 35 *Energy Policy* 2683, 2687–8.

[101] *Renewable Energy Act* 2011, Arts 1(1)–(2) (Ghana).

TABLE 4.18 *Countries Citing 'Promote the Development of the Internal Energy Market and Regional Integration' in Their Legislative Objectives*

Rank of objective	Countries
1	Croatia, Italy, The Netherlands, Paraguay, Sweden.
4	Austria.
5	Lichtenstein.
6	Czech Republic, Greece, Macedonia, Portugal.
7	Kosovo, Montenegro.
10	Albania, Morocco, Serbia.

This focus on the internal market and regional integration reflects the EU's 2030 and 2050 Energy Strategies,[102] as detailed in the documentation on the Energy Union,[103] the European Energy Security Strategy,[104] and Clean Energy for all Europeans.[105] The key dimensions of the Energy Union are:

- 'security, solidarity and trust: diversifying Europe's sources of energy and ensuring energy security through solidarity and cooperation between EU countries;
- a fully integrated internal energy market: enabling the free flow of energy through the EU through adequate infrastructure and without technical or regulatory barriers;
- energy efficiency: improved energy efficiency will reduce dependence on energy imports, lower emissions, and drive jobs and growth;
- decarbonising the economy: the EU is committed to a quick ratification of the Paris Agreement and to retaining its leadership in the area of renewable energy; and

[102] European Commission, 'Communication from the Commission to the European Parliament, the Council, the European Economic and Social Committee and the Committee of the Regions, A policy framework for climate and energy in the period from 2020 to 2030' COM (2014) 15 final; European Commission, 'Communication from The Commission to the European Parliament, the Council, the European Economic and Social Committee and the Committee of the Regions, Energy Roadmap 2050' COM(2011) 885 final.

[103] European Commission, 'Communication from the Commission to the European Parliament, the Council, the European Economic and Social Committee, the Committee of the Regions and the European Investment Bank A Framework Strategy for a Resilient Energy Union with a Forward-Looking Climate Change Policy' COM(2015) 80.

[104] European Commission, 'Communication from the Commission to the European Parliament and the Council, European Energy Security Strategy' COM(2014) 330 final.

[105] European Commission, 'Communication from the Commission to the European Parliament, the Council, the European Economic and Social Committee, the Committee of the Regions and the European Investment Bank Clean Energy for All Europeans' COM (2016) 860 final.

- research, innovation and competitiveness: supporting breakthroughs in low-carbon and clean energy technologies by prioritising research and innovation to drive the energy transition and improve competitiveness'.[106]

It is particularly important for the four EU candidate countries and the potential EU candidate country, Kosovo, to align their laws with these objectives in mind to help smooth the accession process for their domestic and commercial electricity consumers. This concern is particularly apparent from Montenegro's law, which states that the purpose of the law is:

- to link the Montenegrin energy system with the European energy systems and the systems of neighboring countries, in line with energy needs and the needs of economic development;
- to develop the energy market and to link it with the regional and local market of the European Union.[107]

The inclusion of this objective in the Moroccan legislation reflects their desire to gain access to the energy markets of the EU. This latter point is made apparent from the Preamble to the Moroccan legislation that states that one of the main aims of the law is to strengthen 'regional integration through the opening of energy to the Euro-Mediterranean markets and the harmonisation of energy laws and regulations'.[108] The rationale for including the objective in the Paraguayan legislation is similar and relates to the fact that it is a net energy exporter, exporting hydropower generation from two jointly operated large hydropower facilities, Itaipu (operated with Brazil) and Yacyretá (operated with Argentina).

4.4.5.2 Meet International Treaty Obligations and International Agreements

Greenhouse gas emissions and climate change are global problems, which have resulted in international action to try to address these issues. As stated previously, a number of international treaties and agreements affect the renewable energy sector. Despite this, only eleven countries sought to use their renewable energy laws to help them meet their international treaty obligations and international agreements (see Table 4.19).

106 European Commission, *Building the Energy Union* (European Commission, 2018) <https://ec.europa.eu/energy/en/topics/energy-strategy-and-energy-union/building-energy-union>.

107 *Zakon o energetici* [Energy Law, Law as published in the Official Gazette of Montenegro, No. 5/2016 and 51/2017, Art 8] [Montenegro Investment Promotion Agency translation from Montenegrin].

108 *Loi n° 13–09 relative aux énergies renouvelables* [Renewable Energy Law] (Kingdom of Morocco) 2010, Preamble [Tallulah Bur translation from French].

TABLE 4.19 *Countries Citing 'Meet International Treaty Obligations and International Agreements' in Their Legislative Objectives*

Rank of objective	Countries
1	Greece, Peru.
2	Sweden.
4	Lithuania, Pakistan.
6	Mauritius.
7	Albania, El Salvador.
9	Madagascar.
12	Kazakhstan.

The average weighted rank of this category of legislative objective is 5.55. This legislative objective is often given a low priority, possibly because many countries directly incorporate the principles of international treaties and agreements and thus do not need to place further emphasis within their domestic renewable energy law. Another alternative explanation may be that countries cover off this objective through the use of other categories of legislative objectives such as 'reduce greenhouse gas emissions and address climate change'.

An example of this legislative objective is found in the Preamble to the Peruvian law, which states that the aim of the law is to

> facilitate the implementation of the Peru-United States Trade Promotion Agreement and its Protocol of Amendment.[109]

4.4.6 *Environmental Objectives*

As shown previously in Table 4.2, environmental objectives are the third most cited justification for legislating to accelerate or promote the deployment of renewable energy. While arguably many of the categories of legislative objective in this theme are seeking to address the market failures resulting from unpriced negative environmental externalities associated with fossil fuel use and the positive externalities associated with renewable generation, this is not reflected in the overall weighted rank. The weighted

[109] *Decreto Legislativo De Promoción De La Inversión Para La Generación De Electricidad Con El Isuo De Energías Renovables* [Legislative Decree of Investment Promotion for Electricity Generation with the Use of Renewable Energy (Peru) No. 1002/2008, Preamble [Linguistico Translations translation from Spanish].

average rank for environmental objectives is 5.38, which means it is assigned a lower priority than other themes such as sectoral objectives that do not involve a market failure.

4.4.6.1 Environmental Protection

Of all the categories of legislative objectives, the category of legislative objective targeting 'environmental protection' was the most common, having been adopted by fifty-five countries. This is nearly half of the 113 countries with renewable energy laws. This objective seeks to address many of the environmental problems associated with conventional fossil fuel use. This may include problems such as air pollution, unsustainable water use and pollution, thermal pollution and waste, as well as their consequences such as increased health problems, acid rain, waste disposal and loss of biodiversity.

The countries that legislate with the objective of 'environmental protection' (see Table 4.20), on average have significantly higher levels of energy self-sufficiency (92 per cent) than the average for those countries with renewable energy laws but without that objective (76 per cent).[110] On average, countries with the objective also have higher levels of electric power consumption (4,832kWh/per capita), than countries with renewable energy laws but without the objective (3,926kWh/per capita).[111] Despite the range of problems that countries that target 'environmental protection' seek to address, Lyon and Yin, in their empirical study of the adoption of Renewable Portfolio Standards in the United States, found that 'local environmental conditions and preferences ha[d] no significant effect on the timing of adoption'.[112]

When this objective is prioritised, it is important to recognise the specific problems that the legislation is seeking to address. This may not only have an impact on the fossil fuel generators but may also affect the renewable energy sector. For example, if legislators are particularly concerned with air pollution, this may mean that power generated from photovoltaic solar, wind and hydropower is given preference over power generated from woody biomass. Equally, where legislators are concerned about water usage, this may steer energy project developers away from not

[110] See above n 45.

[111] Author's own calculations based on World Bank data on electric power consumption. (Source: The World Bank, *Electric Power Consumption kWh/per capita* (2018) <databank.worldbank.org/data/reports.aspx?source=2&series=EG.USE.ELEC.KH.PC>).

[112] Thomas P Lyon and Haitao Yin, 'Why Do States Adopt Renewable Portfolio Standards? An Empirical Investigation' (2010) 31 *The Energy Journal* 131.

TABLE 4.20 *Countries Citing 'Environmental Protection' in Their Legislative Objectives*

Rank of objective	Countries
1	Algeria, South Korea, Luxembourg.
2	Australia, Austria, Cameroon, Czech Republic, Ghana, Kazakhstan, Panama, Taiwan, Venezuela.
3	The Bahamas, Belarus, Estonia, Germany, Jordan, Tajikistan, Yemen.
4	Lichtenstein, Montenegro, Suriname, Togo.
5	China, Croatia, Dominican Republic, Japan, Kyrgyzstan, Lithuania, The Netherlands, Palestine, Switzerland, Ukraine.
6	Albania, Colombia, Iceland, Kosovo, Turkey.
7	Ecuador, The Gambia, Peru, Romania, South Africa.
8	Democratic Republic of the Congo, Morocco, Paraguay, Serbia.
9	Macedonia, FYR, Russia, Thailand.
10	France.
11	Bulgaria, Philippines.
14	Indonesia.

just coal and nuclear-fired generation but also away from large biomass, concentrated solar thermal and geothermal generation, all of which are heavily water intensive/per kWh generated. Further, and as discussed in Chapter 2, it is environmental concerns that have also prompted some countries to remove large-scale hydropower from their supported renewable energy sources.

Examples of legislative objectives targeting 'environmental protection' include:

- to mitigate the adverse environmental effects of energy operations using fossil fuels[113]
- The purpose of this Act is to enable the energy supply to develop in a sustainable manner . . . in the interest of mitigating climate change and protecting the environment . . .[114]

113 *Ley No. 5707 sobre Incentivo al Desarrollo de Fuentes Renovables de Energía y de sus Regímenes Especiales* [Renewable Energies Incentive Law 57-07] (Dominican Republic) 2007 [Linguistico Translations translation from Spanish].

114 *Gesetz für den Ausbau erneuerbarer Energien (Erneuerbare-Energien-Gesetz – EEG 2017)* [Renewable Energy Sources Act 2017], s 1(1)[German Federal Ministry for Economic Affairs and Energy translation from German].

- The promotion of renewable energies aims to: protect the environment by promoting the use of non-polluting sources of energy[115]
- The objects of this Act are: [. . .] (c) to ensure that renewable energy sources are ecologically sustainable.[116]
- The purpose of this Act is to contribute to the preservation of the environment . . .[117]

4.4.6.2 Reduce Greenhouse Gas Emissions and Address Climate Change

Given the current reliance on fossil fuel generation in many countries, there is a strong causal link between energy use and the production of greenhouse gas emissions. Indeed, energy use is responsible for 66 per cent of total greenhouse gas emissions and 80 per cent of carbon dioxide emissions.[118] Further, the GHG emissions produced from energy use are increasing at a much faster rate than the GHG emissions from other sources, particularly in developing countries.[119] This is problematic as research has shown that the larger a country's carbon dioxide emissions, the smaller their commitment to renewable energy tends to be.[120] Marques et al. have stated that 'this suggests that the greater the level of economic activity, the greater the pollutant activity will be and therefore the propensity to invest in renewable sources will be smaller'.[121]

Legislative objectives targeted at reducing greenhouse gas emissions and addressing climate change are commonly included to encourage compliance with the country's obligations under the international climate change treaties. In some instances, these objectives may also be seeking to address a perceived

[115] *Loi n° 04–09 du 27 Joumada Ethania 1425 correspondant au 14 août 2004 relative à la promotion des énergies renouvelables dans le cadre du développement durable* [Law No. 04–09 of 27 Jumada Ethania 1425 corresponding to 14 August 2004 on the promotion of renewable energies in the framework of sustainable development] (Algeria) [Stephanie Watson translation from French].

[116] *Renewable Energy (Electricity) Act* 2000 (Cth) s 3.

[117] 신에너지 및 재생에너지 개발・이용・보급 촉진법 [Act on the Promotion of the Development, Use and Diffusion of New and Renewable Energy, No. 14670, 2017] [Korean Legislative Research Institute translation from Korean].

[118] IEA, *Climate Change* (OECD/IEA, 2018) <www.iea.org/topics/climatechange/>.

[119] IEA, (March 2018) *Global Energy and CO2 Status Report* 2017 (OECD/IEA, 2017) 3–4.

[120] António C Marques, José A Fuinhas and J R Pires Manso, 'Motivations Driving Renewable Energy in European Countries: A Panel Data Approach' (2010) 38 *Energy Policy* 6877, 6883. This still holds true today as evidenced by a comparison of the CO_2 emissions in metric tons per capita (World Bank databank) and the national renewable power targets in place at the end of 2017.

[121] Ibid.

economic risk associated with future climate agreements or regulations.[122] An unexpected outcome of this research was that the level of concern that countries express about climate change and greenhouse gas emissions through international agreements such as the United Nations Framework Convention on Climate Change[123] and the Paris Agreement[124] was not reflected in the number of countries targeting it as a legislative objective in their renewable energy laws. For example, while 197 countries are a party to the Paris Agreement, only 28 countries have a legislative objective of reducing greenhouse gas emissions and addressing climate change (see Table 4.21). The striking feature shared by many of the countries that do have this as a legislative objective is that they were very early parties to one of the predecessors to the Paris Agreement, the Kyoto Protocol,[125] with most of these countries agreeing to it in the period of 1999 to 2002. Despite this, it is often not a highly weighted legislative objective, with its average weighted rank being only 5.89.

TABLE 4.21 *Countries Citing 'Reduce Greenhouse Gas Emissions and Address Climate Change' in Their Legislative Objectives*

Rank of objective	Countries
1	Australia, Austria, Bangladesh, Czech Republic.
2	Algeria, Germany.
3	Denmark, World (IRENA).
4	Mexico, Tajikistan.
5	Albania, The Gambia, Pakistan, Togo, Turkey.
6	El Salvador, Estonia.
7	Greece.
8	Romania, Senegal.
9	France, South Korea, Morocco, Philippines.
10	Macedonia, FYR.
11	Kazakhstan.
12	Bulgaria.
14	Peru.

122 Grace et al., above n 36, 3.

123 *United Nations Framework Convention on Climate Change*, opened for signature 4 June 1992, 1771 UNTS 107 (entered into force 21 March 1994).

124 *The Paris Agreement*, opened for signature 16 February 2016, Decision 1/CP.21, Annex, UN Doc FCCC/CP/2015/10/Add.1 (29 January 2016) (entered into force 4 November 2016).

125 *Kyoto Protocol to the United Nations Framework Convention on Climate Change*, opened for signature 11 December 1997, 2303 UNTS 148 (entered into force 16 February 2005).

This may be because some economists argue that regulatory support mechanisms for renewable energy distort emissions trading schemes (ETS) such as the EU ETS by lowering demand for fossil fuels and hence the effective price of carbon.[126] They argue that specific renewable energy policies should not be used if their purpose is to reduce greenhouse gas emissions and mitigate climate change.[127] Alternatively, they may be concerned about the potential interaction between renewable energy laws and climate change laws.[128] However, as can be seen from both the previous chapter on the economic justification for regulatory intervention and the current chapter, most countries are trying to achieve much more than reducing greenhouse gas emissions through their renewable energy laws. Further, those countries that do seek to incorporate it as a legislative objective on average place quite a low priority on it relative to other objectives such as energy security, diversifying supply, sustainable development, strengthening the economy or even improving system stability, safety and reliability.

Countries that do prioritise this objective are often focused on rapidly deploying renewable generation, regardless of the size or location of the project. They may also prefer to support renewable energy technologies that produce lower greenhouse gas emissions over their lifecycle. This may mean that woody biomass and new large-scale hydropower plants may be avoided due to their effect on the release of methane, whereas landfill and sewage gas generation may be promoted due to their benefits in using methane that would otherwise be released into the atmosphere.[129]

Examples of objectives targeting the reduction of greenhouse gas emissions and addressing climate change are found in both the Preamble to the IRENA Statute and in the Algerian law respectively:

- 'Convinced of the major role that renewable energy can play in reducing greenhouse gas concentrations in the atmosphere, thereby contributing to the stabilisation of the climate system, and allowing for a sustainable, secure and gentle transit to a low carbon economy.'[130]

126 Carolyn Fischer and Richard G Newell, 'Environmental and Technology Policies for Climate Mitigation' (2008) 55 *Journal of Environmental Economics and Management* 142; Max Rathmann, 'Do Support Systems for RES-E Reduce EU-ETS-Driven Electricity Prices?' (2007) 35 *Energy Policy* 342.

127 Philibert, above n 93, 5.

128 Fischer et al., above n 41.

129 Grace et al., above n 36, 17.

130 International Renewable Energy Agency, *Statute of the International Renewable Energy Agency* (Adopted at the Conference on the Establishment of the International Renewable Energy Agency, Bonn, 26 January 2009, entered into force 8 July 2010), Preamble.

- 'The promotion of renewable energies aims to: [. . .] contribute to the fight against global warming by reducing emissions of greenhouse gases.'[131]

4.4.6.3 Reduce the Risk of Natural and Nuclear Disasters

The legislative objective of reducing the risk of natural and nuclear disasters only features in the Bangladeshi and the French laws, with a weighted rank of 6.50. This is likely to reflect the fact that Bangladesh is highly prone to natural disasters such as severe flooding, tropical cyclones, earthquakes and landslides. A considerable body of research has established that the frequency of natural disasters has increased with global warming and climate change.[132] Therefore, by encouraging a shift from fossil fuel generation to renewable generation, the idea is that this will in turn reduce the risk of natural disasters. This objective is contained in the Preamble to the Sustainable and Renewable Energy Development Authority Act 2012 (Bangladesh), which states that one of the purposes of the Act is 'to mitigate the risk of natural calamity'.

The rationale for France inserting the objective was somewhat different. France introduced their Energy Transition Law in August 2015. However, the commitment to significantly reduce France's nuclear generation occurred in the immediate aftermath to the Fukushima nuclear disaster and was the outcome of political negotiations for the alliance between President Hollande and the Green Party during the 2012 Presidential elections.[133] The drafting of the law and the ensuing parliamentary debate led to intense debate about the future of nuclear generation in the French energy mix. As a result, France's law reflects a concern to 'guarantee nuclear safety', by avoiding nuclear accidents and the safe disposal of nuclear waste at the end of life.

[131] *Loi n° 04–09 du 27 Joumada Ethania 1425 correspondant au 14 août 2004 relative à la promotion des énergies renouvelables dans le cadre du développement durable* [Law No. 04–09 of 27 Jumada Ethania 1425 corresponding to 14 August 2004 on the promotion of renewable energies in the framework of sustainable development], Art 2 (Algeria) [Stephanie Watson translation from French].

[132] See e.g. Intergovernmental Panel on Climate Change, *Managing the Risks of Extreme Events and Disasters to Advance Climate Change Adaptation (SREX)* (2012) IPCC <www.ipcc.ch/pdf/special-reports/srex/SREX_Full_Report.pdf>.

[133] Michael Stothard, 'France's nuclear industry on back foot over new energy law', *Financial Times*, 27 November 2015 < www.ft.com/content/489bea54-84a3-11e5-8095-ed1a37d1e096>.

4.4.7 *Industrial Policy Objectives*

We're in a competition all around the world, and other countries – Germany, China, South Korea – they know that clean energy technology is what is going to help spur job creation and economic growth for years to come. And that's why we've got to make sure that we win that competition. I don't want the new breakthrough technologies and the new manufacturing taking place in China and India. I want all those new jobs right here . . . in the United States of America, with American workers, American know-how, American ingenuity.[134] – President Barack Obama

The renewable energy sector has become a fertile ground for countries to exercise industrial policy. To this end, countries provide regulatory support to the renewable energy sector to assist in the development of local manufacturing sectors in order to become leading producers of renewable energy technologies. In this way, countries often also tried to stimulate jobs and foster technological innovation, thereby encouraging the development of new industries and infrastructure within a country and the development of an export market.[135]

Experience from countries such as Germany and Japan has shown that while industry stimulus may be successful in the short-term, in the face of growing competition from China, other countries often struggle to sustain the growth into the longer term.

4.4.7.1 Support the Development of New Industry and Infrastructure

Twenty-four countries 'support the development of new industry and infrastructure' through their renewable energy laws (see Table 4.22), with it being assigned an average weighted rank of 5.38. These countries include the industrial powerhouses of China and South Korea, as well as a number of emerging economies, such as South Africa, Indonesia and the Philippines. In 2016, the average GDP per capita of the countries that had adopted this legislative objective was $US9,995 compared to an average of $US16,024 for all countries with a renewable energy law.[136]

Legislative objectives that target the development of new industries and infrastructure operate on the basis that the particular industries or infrastructure are so important to the country that they justify distorting the market through regulation supporting them. These objectives may be the result of

[134] Barack Obama (Remarks, Allison Transmission Headquarters, Indiana) cited in Adele C Morris, Peitro S Nivola and Charles L Schultze, 'Clean Energy: Revisiting the Challenges of Industrial Policy' (2012) 34 *Energy Economics* S34, S34.

[135] Morris et al., above n 134, S36.

[136] Author's own calculations based on World Bank data on GDP per capita 2016 (current $US) rounded to the nearest whole dollar. (Source: The World Bank, *World Development Indicators*, above n 57.)

TABLE 4.22 *Countries Citing 'Support the Development of New Industry and Infrastructure' in Their Legislative Objectives*

Rank of objective	Countries
1	China, Honduras, Nicaragua, Portugal.
2	Togo.
3	Lithuania, Taiwan, Tonga.
5	Armenia, Democratic Republic of the Congo, Ecuador.
6	Austria, Bulgaria, Cuba, Kazakhstan, Paraguay.
7	Croatia, Hungary, South Korea, Senegal.
8	The Bahamas.
9	South Africa.
12	Indonesia, Philippines.

political compromise and may ensure that regulatory support goes to industries prevalent in specific geographic areas within a country.

An example of how legislative objectives target this area is found in the South Korean legislation:

> The purpose of this act is to contribute to [...] use and distribution of new and renewable energy, and the activation of the new energy industry and renewable energy industry ...[137]

4.4.7.2 Encourage Technological Innovations

The rationale behind encouraging technological innovation is linked to the previous legislative objective, that is, that technological innovation may facilitate new industry and export markets in the country. In addition, by encouraging the development and subsequent commercialisation of emerging technologies, it may make electricity generation from renewable sources more efficient and in the long term more likely to achieve cost competitiveness with fossil fuel sources.[138] Other benefits include the potential to reduce some of the costs and risks currently associated with technologies in the sector, while increasing the benefits. Huang et al. have also argued that technological innovations can lead to expanded energy supplies, and improve the availabil-

[137] 신에너지 및 재생에너지 개발・이용・보급 촉진법 [Act on the Promotion of the Development, Use and Diffusion of New and Renewable Energy, No. 14670, 2017] [Korean Legislative Research Institute translation from Korean].

[138] Grace et al., above n 36, 3.

ity and quality of energy, all while reducing the adverse environmental impacts that result from energy extraction, conservation and usage'.[139]

Technological innovation was supported by twenty-one countries (see Table 4.23) through their renewable energy laws, with an average weighted rank of 4.95:

TABLE 4.23 *Countries Citing 'Encourage Technological Innovations' in Their Legislative Objectives*

Rank of objective	Countries
1	Uzbekistan.
2	Argentina, Lithuania, Russia, Tonga.
3	Albania, Madagascar.
4	San Marino, Slovenia, Venezuela, World (IRENA).
5	Bulgaria, Paraguay.
6	South Korea, Peru, Senegal.
7	Germany, Palestine.
8	South Africa, Ukraine.
9	Croatia.
11	Morocco.

Countries that support technological innovation are likely to design their regulatory support mechanisms in such a way as to band the support of renewable energy technologies, rather than simply preferring the technology that is currently least-cost. The banding of technologies may be done in several ways, including carving out projects and technologies that are already cost competitive or well-established, and applying a multiplier to the support given to emerging technologies based on their level of commercialisation. Ideally, renewable energy laws should enable regulatory support mechanisms to be provided for all phases of the innovation cycle from research and development to commercialisation to provide legal certainty and market stability.[140]

An example of this objective is found in the Lithuanian law:

> This Law aims to ensure sustainable growth in the exploitation of renewable energy sources as well as to promote further development and implementation of relevant new technologies . . .[141]

139 Cui Huang et al., 'Government Funded Renewable Energy Innovation in China' (2012) 51 *Energy Policy* 121, 121.

140 Jeffrey M Loiter and Vicki Norberg-Bohm, 'Technology Policy and Renewable Energy: Public Roles in the Development of New Energy Technologies' (1999) 27 *Energy Policy* 85, 95; See also Davies, 'Reconciling Renewable Portfolio Standards and Feed-in Tariffs', above n 4, 321.

141 *Atsinaujinančių išteklių energetikos įstatymas* [Law on Energy from Renewable Sources No. XI-1375, 2011], part 1 Art 1(2) [Linguistico Translations translation from Lithuanian].

4.4.7.3 Create Jobs or Improve Skills and Domestic Capabilities

Legislative intervention in the renewable energy sector is sometimes justified on the basis that regulatory support mechanisms stimulate job creation. As can be seen in Table 4.24, job creation and improving domestic capabilities tend to be a focus of lower to upper-middle income countries, though it is often not a high priority, with an average weighted rank of 6.23. Further, Lyon and Yin's empirical study of the adoption of Renewable Energy Portfolio Standards in the United States found that the states with high unemployment rates were slower to adopt a renewable portfolio standard (RPS) than those with lower unemployment rates.[142] On average, the twelve countries that cited job creation in their legislative objectives had an annual GDP per capita of $US8,756 in 2016, which is roughly half that for all countries with a renewable energy law.[143]

An example of this legislative objective is found in the Romanian law:

> This law creates a legal framework necessary to broaden the use of renewable energy by: (b) stimulating sustainable development at the local and regional level and creating new jobs associated with recovery processes of renewable energy.[144]

TABLE 4.24 *Countries Citing 'Create Jobs or Improve Skills and Domestic Capabilities' in Their Legislative Objectives*

Rank of objective	Countries
3	France.
4	Honduras, Kazakhstan.
5	Portugal.
6	Croatia, Dominican Republic, Romania, Togo, Tonga, World (IRENA).
8	The Gambia, Ghana.
13	Indonesia.

142 Lyon et al., above n 112.

143 Author's own calculations based on World Bank data on GDP per capita 2016 (current $US) rounded to the nearest whole dollar. (Source: The World Bank, *World Development Indicators*, above n 57.)

144 *Legea 220/2008 pentru stabilirea sistemului de promovare a producerii energiei din surse regenerabile de energie, republicata 2010* [Law 220/2008 on establishing the promotion system of energy production from renewable energy sources] (Romania) ch 1 Art 1(1) [Linguistico Translations translation from Romanian].

IRENA has reported that in 2017 10.3 million people were directly and indirectly employed within the renewable energy sector globally.[145] This is predicted to rise to 23.6 million people in 2030 and 28.8 million people in 2050.[146] However, the research shows that to date employment within the renewable energy sector has been highly concentrated in a just few countries that have actively competed to dominate the sector in terms of technological innovation and the manufacturing of equipment. For example, 43 per cent of all of the renewable energy jobs globally are currently located in China, with Brazil, the United States, India, Germany and Japan also having significant numbers of people employed within the sector.[147] These figures look even starker when analysed by technology, with just five countries responsible for 90 per cent of the global employment in the photovoltaic solar sector globally.[148] Further, China is solely responsible for 2.2 million or 66 per cent of the global photovoltaic solar jobs. What is striking about this result is that none of these countries have identified job creation or the improvement of skills and domestic capabilities as one of their legislative objectives.

In contrast, only 76,000 renewable sector jobs were located on the African continent in 2017,[149] which is fewer than the number of people employed within the sector in Turkey.[150] This highlights the challenges of attempting to create jobs within the sector. For jobs to be created in any large numbers there needs to be access to high tech manufacturing facilities, a skilled workforce and also, ideally, the ability to scale production to reduce costs. If a country does not have this capacity, the most it can hope for in creating jobs is to create jobs in the construction, operation and maintenance of renewable energy facilities. However, the jobs in the construction phase of a facility are temporary, and there tend to be comparatively fewer jobs in the operation and maintenance phases of many renewable energy facilities such as large scale photovoltaic solar.

As the Chinese industrial and technological dominance within the renewables sector has grown, there have been corresponding job losses within the sector in other parts of the world. For example, there has been an 11 per cent decline in jobs within the PV solar industry in Japan, an 8 per cent decline in the European Union and 3.8 per cent decline in the United States.[151] The employment figures

145 IRENA, *Renewable Energy and Jobs Annual Review 2018* (IRENA, 2018) 3.
146 Ibid 24.
147 Ibid 3.
148 Ibid 7.
149 Ibid.
150 Ibid 3 and 21.
151 Ibid 7.

also do not consider the net employment impact of providing regulatory support to the renewable energy sector; in particular, the projected job losses in fossil fuel fired generation, and associated industries. A further area of concern involves indirect job losses due to higher electricity prices while the sector transitions, which may cause highly energy intensive industries such as aluminium smelting to shut down and/or to move abroad if they are not given preferential treatment. Indeed, Frondel et al. have argued that, contrary to the perception that renewable energy legislation can be used to create jobs,

> numerous empirical studies have consistently shown the net employment balance to be zero or even negative in the long run, a consequence of the high opportunity cost of supporting renewable energy technologies. Indeed, it is most likely that whatever jobs are created by renewable energy promotion would vanish as soon as government support is terminated, leaving only Germany's export sector to benefit from the possible continuation of renewable support in other countries such as the US.[152]

Similar results have been achieved in other studies of the employment effects of renewable energy laws.[153] For example, in Australia when the available feed-in tariff under the NSW Solar Bonus Scheme was reduced and then new applications under the scheme were closed, it had an almost immediate effect on employment levels and business closures within the NSW solar industry. A survey conducted by the Australian Solar Energy Society in August 2011 of ninety-one solar businesses in NSW stated that there had been 'a 93% fall in sales inquiries and more than 400 job losses' since November 2010.[154] Further, '25% of the businesses contacted were either closed or planning to close'.[155] This would suggest that the poor design of the NSW Scheme actively hampered its second objective of 'developing jobs in the renewable energy sector by assisting renewable energy generation to compete with non-renewable energy generation',[156] as the jobs created were not sustainable without the assistance of the feed-in tariff.

152 Frondel, above n 96, 4055.

153 Morris et al., above n 134, S39.

154 Amos Aikman, 'Government's Withdrawal of Solar Subsidy Scheme Leaves Industry in Trouble', *The Australian* (online), 18 August 2011 <www.theaustralian.com.au/national-affairs/governments-withdrawal-of-solar-subsidy-scheme-leaves-industry-in-trouble/news-story/d8c883711e267a2f3d55620d872744e6?sv=eb2c60eb7cce9a9b59e85904e96112fd>.

155 Ibid. Note, it is unclear how much blame for the reduction in employment opportunities within the sector can be attributed to the removal of the NSW Solar Bonus Scheme given the other ancillary pressures present in the market at the time, including the effects of the global financial crisis and general market turmoil.

156 *Electricity Supply Amendment (Solar Bonus Scheme) Act 2009* (NSW) sch 1 s 15(1A)(1)(b), which amended the *Electricity Supply Act 1995* (NSW).

4.4.7.4 Local Manufacturing

The rationale behind using renewable energy to support local manufacturing is that it can 'ameliorate unemployment, stimulate growth, and stem the 'offshoring' of manufacturing',[157] leading to increased economic growth and national income in the long-term. That said, only five countries (see Table 4.25) explicitly have as an objective within their renewable energy laws to support local manufacturing:

TABLE 4.25 *Countries Citing 'Local Manufacturing' in Their Legislative Objectives*

Rank of objective	Countries
2	Peru.
3	Argentina.
6	Palestine.
7	Turkey.
8	Japan.

This is likely to be for two reasons. First, it may be that countries do not believe that their renewable energy laws are the most efficient and effective vehicle for achieving this objective.[158] Second, the reluctance to include local manufacturing as a legislative objective may also reflect concerns about domestic content clauses under the Agreement on Subsidies and Countervailing Measures (SCM) to the WTO.[159] Despite this, Turkey currently provides an additional incentive of between $US0.004 and 0.024/kWh for five years for renewable energy facilities that use Turkish-manufactured mechanical and/or electromechanical equipment and are operational prior to 31 December 2020.[160] Local content requirements are also a common feature of Turkish auctions for large-scale solar and offshore wind projects, which

157 Morris et al., above n 134, S36.

158 See e.g. OECD, *OECD Policy Guidance for Investment in Clean Energy Infrastructure* (OECD, 2013) 20.

159 *Marrakesh Agreement Establishing the World Trade Organisation*, opened for signature 15 April 1994, 1867 UNTS 3 (entered into force 1 January 1995 annex 1A ('*Subsidies and Countervailing Measures*') Arts 3.1–2. See e.g. Vyoma Jha, 'Political Economy of Climate, Trade and Solar Energy in India' (2018) IX(2) *Trade, Law and Development* 138.

160 Council of Ministers Decision No. 2013/5625 published in the Official Gazette No. 28842, dated 5 December 2013 (Turkey).

often require at least 60 per cent of the equipment to be domestically manufactured.[161]

An example of this category of legislative objective is found in the Turkish law:

> The purpose of this Law is to expand the use of renewable energy sources for the purpose of energy generation, to introduce such sources into the economy in a reliable and economic fashion and to a good quality standard, to broaden the variety of sources, to reduce greenhouse gas emissions, to exploit waste, to protect the environment, and to develop the manufacturing sector as required in order to fulfil these purposes.[162]

4.4.8 *Social Objectives*

The four categories of legislative objectives within the social theme are:

1. public health or improving living standards or social development;
2. improved access to electricity;
3. affordable energy; and
4. promote rural development.

The social objectives are often not highly prioritised, with an average weighted rank of 5.98, making it the lowest ranking theme out of all of the eight themes.

4.4.8.1 Public Health or Improving Living Standards or Social Development

This legislative objective has two central but interrelated subjects. First, when it is used to target public health, it commonly refers to reducing reliance on fossil fuel generation or traditional biomass use, which makes up much of the domestic fuel used in the developing world.[163] These fuels have been associated with particulate air pollution and negative health impacts such as increased risk of heart attacks, strokes, lung disease and asthma. They also have serious impacts on child mortality, with the World Health Organization (WHO) reporting that 4 million premature deaths occur each year as a result

161 David Foxwell, 'Local Content a Key to Upcoming Offshore Wind Tender', *Offshore Wind Journal*, 20 June 2018 <www.owjonline.com/news/view,local-content-a-key-to-upcoming-turkish-offshore-wind-tender_52205.htm>.

162 *Yenilenebilir Enerji Kaynaklarinin Elektrik Enerjisi Üretimi Amaçli Kullanimina İlişkin Kanun Kanun Numarası* [Law Regarding The Use Of Renewable Energy Resources For Electricity Production] (Turkey) 2005 No. 5346, s 1 [Linguistico Translations translation from Turkish].

163 Tariq Banuri and Niclas Hällström, 'A Global Programme to Enable Energy Access and Climate Change' (2012) 3 *What Next* 264, 268.

of indoor smoke inhalation, and more than half of these deaths occur among children less than five years of age.[164] Second, the use of traditional fuels such as firewood, charcoal and animal dung has a marked impact on the living standards and social development of communities,[165] and has a particularly negative impact upon the women in the community, who are often responsible for collecting the firewood and animal waste.[166]

The average weighted rank among the nineteen countries that cite this objective is 7.11 (see Table 4.26).

TABLE 4.26 *Countries Citing 'Public Health or Improving Living Standards or Social Development' in Their Legislative Objectives*

Rank of objective	Countries
1	Democratic Republic of the Congo.
2	El Salvador, Pakistan.
3	Venezuela.
4	Belarus.
5	The Bahamas, Slovenia.
6	South Africa, Ukraine.
7	Armenia, Honduras.
8	France.
10	Croatia, Japan, Philippines, World (IRENA).
11	Indonesia.
13	Bulgaria.
15	Peru.

Examples of this legislative objective include:

- 'improving the living standards of the population through economically efficient use of energy from renewable sources'.[167]
- 'Preserve human health and the environment, in particular by combating the aggravation of the greenhouse effect and against major industrial

164 World Health Organisation, 'Household Air Pollution and Health' (Fact Sheet WHO, 8 May 2018).

165 Kamil Kaygusuz, 'Energy Services and Energy Poverty for Sustainable Rural Development' (2011) 15 *Renewable and Sustainable Energy Reviews* 936.

166 See also Wenguang Ding et al., 'Impacts of Renewable Energy on Gender in Rural Communities of North-West China' (2014) 69 *Renewable Energy* 180, 188.

167 Закон за енергията от възобновяеми източници [Energy from Renewable Sources Act] (Republic of Bulgaria) 3 May 2011, State Gazette 35, 2011, ch 1 Art 2.

> risks, by reducing the exposure of citizens to air pollution and by guaranteeing the nuclear safety.'[168]

4.4.8.2 Improved Access to Electricity

Globally, access to electricity is very unevenly distributed with an estimated 1.1 billion people, or approximately 14 per cent of the global population, lacking access to electricity.[169] Research conducted by the Islamic Investment Bank found that '70% of Africans have no access to electricity, while the entire electricity generation capacity of Sub-Saharan Africa is 68% of that of Spain'.[170] This problem is particularly acute in rural communities, with more than 90 per cent of people lacking access to energy being rural dwellers.[171] It is estimated that $US725 billion of investment (or just over three years' worth of global fossil fuel subsidies) would be required to achieve universal access to electricity globally by 2030.[172]

Energy poverty has a wide range of impacts ranging from preventing income generating economic activities that rely on energy, limiting access to transportation, the internet and telecommunications, poorer health and educational outcomes, and a greater domestic burden being placed upon women.[173] In this way, a lack of access to electricity poses a serious barrier to economic and social development.

This problem does not confine itself to domestic consumers, with commercial users in Africa 100 times more likely to experience service interruptions such as blackouts than the applicable US standard.[174] Access to electricity is also an issue for developed countries, with Banuri and Hällström noting that 'measured in kilowatt-hours (kWh) per person per day, the global average consumption of primary energy of the richest countries is even more unequally distributed than per capita income'.[175]

Previous research has suggested that when countries that have limited energy resources legislate to improve access to electricity, they are forced to

168 *Code de l'énergie* [Energy Code] (France), Art L100-1 [Legifrance (Government of France) translation from French].

169 IEA, *WEO-2017 Special Report: Energy Access Outlook* (IEA, 19 October 2017) 11.

170 Gorbuz Gonul, *Islamic Development Bank's Approach for Improved Access to Electricity* (May 2013) Islamic Development Bank <www.isdb.org/irj/go/km/docs/documents/IDBDevelopments/Internet/English/IDB/CM/Publications/Electricity.pdf> 7.

171 Ibid.

172 Anna Zinecker et al., *Getting on Target: Accelerating Energy Access Through Fossil Fuel Subsidy Reform: GSI Report* (International Institute for Sustainable Development, 2018) 5.

173 Ibid 8; see also Ding et al., above n 166, 180–1.

174 Gonul, above n 170, 7.

175 Banuri et al., above n 163, 265.

make decisions about the relative importance of allocating electricity for human welfare purposes versus economic development and industrialisation.[176] As has been highlighted above, in the very poorest countries, almost all of the energy is consumed by household consumption and public health needs,[177] often in the form of traditional biomass.[178] In contrast, in the emerging economies that make up the bulk of countries with this objective, a disproportionate share is dedicated to industrial purposes.[179]

In contrast to the high average GDP per capita for legislative objectives targeting energy affordability, the average GDP per capita for countries that are targeting improved access to electricity is only $US7,012. This places the average GDP of these countries in the middle of the 'upper-middle' income bracket according to the World Bank classification. The countries that have adopted this legislative objective (see Table 4.27) on average provide access to electricity to 84.1 per cent of their population, as compared to an average of 93.3 per cent for all countries with renewable energy laws.[180] Despite this, providing access to energy often performs quite poorly on the prioritisation of legislative objectives, with an average weighted rank of 5.06.

TABLE 4.27 *Countries Citing 'Access to Electricity' in Their Legislative Objectives*

Rank of objective	Countries
1	Ecuador, Pakistan, Suriname
2	Democratic Republic of the Congo, Jamaica.
4	Hungary, Madagascar, Palau, Russia.
5	Morocco.
6	Cote d'Ivoire.
7	Ghana, World (IRENA).
8	Indonesia.
9	France.
10	Tonga.
11	Peru.

176 Ibid 266–7.

177 Musiliu O Oseni, 'Improving Households' Access to Electricity and Energy Consumption Pattern in Nigeria: Renewable Energy Alternative' (2012) 16 *Renewable and Sustainable Energy Reviews* 3967, 3967.

178 Banuri et al., above n 163, 266–7.

179 Ibid.

180 Author's own calculations based on World Bank data on access to electricity and GDP per capita 2016 (current $US) rounded to the nearest whole dollar. (Source: The World Bank, *World Development Indicators*, above n 57.)

An example of this legislative objective is found in the Moroccan law:

> The development of national sources of renewable energies is one of the priorities of the national energy policy, whose main aims are: [. . .] providing general access to energy, by the availability of modern energy for all segments of the population and at competitive prices.[181]

4.4.8.3 Affordable Energy

When countries stipulate that they are targeting the cost of energy in their legislative objectives, they are primarily referring to reducing electricity rates for consumers and reducing the risk of fluctuating energy prices. One of the key benefits of renewable energy is that, unlike coal and gas fired generation, much renewable generation does not require costly feedstocks. This means that once a renewable energy project has been constructed, the majority of the costs associated with the project have already been borne. This makes the cost of supplying renewable generation more predictable and less susceptible to price volatility.[182]

Of all the legislative objectives, the countries that were concerned about affordable energy rather ironically had the highest average GDP per capita of $US20,373 in 2016.[183] These twelve countries (see Table 4.28) did not rank affordability of energy as a particularly high priority, with an average weighted rank of 4.31.

TABLE 4.28 *Countries Citing 'Affordable Energy' in Their Legislative Objectives*

Rank of objective	Countries
2	The Bahamas, Estonia, Suriname, World (IRENA).
3	Latvia.
4	The Gambia, Germany, Switzerland, Thailand.
5	Madagascar.
7	France.
8	Russia.
9	Indonesia.

181 *Loi n° 13–09 relative aux énergies renouvelables* [Renewable Energy Law] (Kingdom of Morocco) 2010, Preamble [Tallulah Bur translation from French].

182 Grace et al., above n 36, 14.

183 Author's own calculations based on World Bank data on GDP per capita 2016 (current $US) rounded to the nearest whole dollar. (Source: The World Bank, *World Development Indicators*, above n 57.)

Where the affordability of energy is a high legislative priority, this implies that the focus of the legislation will be providing renewable generation at least cost. This means that renewable energy technologies that are currently the most competitive are likely to be those supported under the legislation regardless of the size of the project, its location, or even whether it is likely to be the most efficient form of renewable generation in the long term.

A further benefit that renewable energy provides in terms of achieving this legislative objective is the so-called price suppression effect.[184] In competitive wholesale electricity markets, the electricity market operator will call for bids to supply electricity within the applicable window of time. The market operator will select the lowest-priced bids until they have sufficient supply to meet the likely electricity demand within a window. Due to the lack of feedstocks used in much of the renewable generation, renewable generators are often able to offer their electricity at a lower cost than some fossil fuel generators. This then lowers the 'bid-stack' and reduces reliance on high cost, and often old, fossil fuel generators. Interestingly, the price suppression effect diminishes over time, as future investment decisions factor this in and renewable generation capacity increases.

An example of a legislative objective targeting energy affordability is found in the Indonesian law:

> in order to support sustainable national development and improve national energy security, the management of energy shall be aimed to: [...] (f) Improve accessibility to energy for communities that are less wealthy and/ or that live in remote areas to bring about fair and equitable welfare and prosperity for the people by: (1) providing assistance to increase the availability of energy to societies that cannot afford it; (2) building energy infrastructure to undeveloped areas so as to reduce disparities among regions.[185]

4.4.8.4 Promote Rural Development

This legislative objective is closely related to the previous objective of improving access to electricity. Renewable energy laws often try to stimulate economic activity and create jobs within particular parts or regions of the country. There are a wide range of measures that may be used to promote rural development, including: additional subsidies available for projects based in

184 Clean Energy Council, 'Clean Energy Council Submission' (Issues paper, Renewable Energy Target Review, 2014) 19.

185 *Undang-Undang Republik Indonesia Nomor 30 Tahun 2007 Tentang Energi* [Law of the Republic of Indonesia Number 30 of 2007 About Energy] (Republic of Indonesia) ch 2 Art 3 [Ellen Marie O'Brien translation from Indonesian].

rural areas; mandating a specified amount of local content in each project; and facilitating research and development in rural areas.[186] The ten countries that are targeting rural development in their legislative objectives have an average access to electricity rate of 86.4 per cent among their populations (see Table 4.29). However, the rural electrification rate is often far lower, with less than 1 per cent of the rural population in the Democratic Republic of the Congo having access to electricity.[187] They have an average weighted rank for promoting rural development of 7.45.

TABLE 4.29 *Countries Citing 'Promote Rural Development' in Their Legislative Objectives*

Rank of objective	Countries
3	Iceland, Senegal (43%).
5	Honduras, Romania.
7	Bulgaria.
8	World (IRENA).
9	Democratic Republic of the Congo (less than 1%), Ecuador (94%).
10	Indonesia.
11	Tonga.
12	Albania.

Countries that seek to promote rural development often focus on the provision of distributed generation. In this way, electricity can be brought to rural and remote communities without the need to build new transmission and distribution infrastructure or upgrade any existing infrastructure.[188] Countries that focus on the promotion of rural development through their renewable energy laws also need to consider how the impact of renewable energy may differ from more urban areas. For example, consideration must be given to traditional land use patterns, farming practices and the impact on the local environment.[189]

An example of a legislative objective promoting rural development is found in the Honduran law:

186 Lipp, above n 22, 5485.

187 IEA, WEO-2017 *Special Report: Energy Access Outlook*, above n 169, 114.

188 Warren Leon, *Evaluating the Benefits and Costs of a Renewable Portfolio Standard* (Clean Energy States Alliance, 2012) 8–9.

189 Ibid 22.

> This Law has the principal aim of promoting public and/or private investment in electricity generation projects with national renewable resources, through fulfilment of the following objectives: [. . .] (5) Increase the quality of life for inhabitants in the rural parts of the country through their participation in the benefits brought by energy developments.[190]

4.5 MULTIPLE AND COMPETING LEGISLATIVE OBJECTIVES

One of the challenges with using legislative objectives sections is that they often contain multiple, and at times conflicting, objectives. This research has found that some countries are seeking to achieve up to fifteen legislative objectives within their legislation, with the average number of objectives being more than five. With this number of legislative objectives involved in the average national renewable energy law, conflicting objectives are often inevitable. This is not a new problem, with the competing objectives contained within renewable energy legislation well-recognised in the previous research.[191]

There are a number of implications that follow from renewable energy laws having multiple and conflicting objectives. First, multiple and conflicting objectives can cause confusion and regulatory uncertainty, if it is not explicitly clear which objective is the dominant objective or how to weigh the costs and benefits to determine which should be the dominant objective. This may 'exacerbate investors' perception of political and regulatory uncertainty, operating as a disincentive for investment in the very technologies and projects that policymakers seek to stimulate through renewable energy policy'.[192] Second, conflict often seems to arise between trying to achieve the desired outcome at least-cost versus other environmental, energy security, industrial policy or technology goals,[193] which may require a longer-term view to be taken or the externalities to be priced into decisions around cost. Third, consideration needs to be directed towards the implications of trying to achieve multiple objectives through a single legislative instrument and how they will all interact.[194] In some circumstances, it may be more appropriate for different

190 *Ley de Promocion a la Generacion de Energia Electrica con Recursos Renovables* 2007 [Law for the Promotion of Electricity Generation with Renewable Resources 2007] (Honduras) 70/2007, ch 1 Art 1 [Linguistico Translations translation from Spanish].

191 See e.g. Geoff Kelly, 'Renewable Energy Strategies in England, Australia and New Zealand' (2007) 38 *Geoforum* 326; Fischer and Preonas, above n 41, 53–6; Grace et al., above n 36.

192 Grace et al., above n 36, 2.

193 Ottmar Edenhofer et al., 'On the Economics of Renewable Energy Sources' (2013) 40(1) *Energy Economics* 512, 513. See also Grace et al., above n 36, 4.

194 Philibert, above n 93, 9.

legislative instruments to be used to achieve different goals. This in turn will make it easier to assess the relative costs and benefits justifying legislative intervention and to evaluate the success or failure of that intervention.

Competing legislative objectives may be the result of political compromise,[195] but the implications of ameliorating those conflicts through the statutory interpretation process warrants more consideration than it is often given. For example, hypothetically, if a renewable energy law does not contain a definition of what constitutes renewable energy, and a regulator has to decide whether a large-scale hydroelectric project is eligible for regulatory support under the legislation, the interpretation of the legislative objectives could have a material effect on the outcome. If the country prioritises energy security within its legislative objectives, it may be that the large-scale hydropower project is deemed to be an eligible project. If, however, the country is concerned about water pollution or loss of biodiversity, it may be that the project is not deemed to be eligible. For this reason, legislative objectives that contain multiple and conflicting objectives without any guidance on the dominant objective may render this task near impossible.

The following legislative objectives section from the Japanese law provides an example of the difficulties involved in interpreting the legislative priorities:

> The purpose of this Act is to promote the use of sources of renewable energy as energy sources for electricity by taking special measures in respect of price, time frame, etc. with regard to the procurement of electricity from sources of renewable energy by electricity utilities, taking into consideration that the use of sources of renewable energy as energy sources is important in securing a stable and appropriate supply of energy appropriate for the economic and social environment in Japan and abroad and in reducing the burden on the environment arising from energy supply, thereby contributing to the strengthening of the international competitiveness of Japan and the sound development of the national economy, including the promotion of Japanese industry and the revitalization of local communities.[196]

4.5.1 *Resolving Conflict Between Competing Objectives*

Where conflict between competing objectives occurs, the common law court will seek to resolve the conflict by applying the principle of harmonious

195 *Carr v. Western Australia* (2007) 232 CLR 138, [5]–[6].

196 電気事業者による再生可能エネルギー電気の調達に関 する特別措置法 [Act on Special Measures Concerning Procurement of Electricity from Renewable Energy Sources by Electricity Utilities]（平成二十三年八月三十日法律第百八号）(Act No. 108 of 30 August 2011), Art 1 [Translation by the Japanese Ministry of Justice].

construction. This principle is premised on two presumptions: one against internal conflict and another in support of coherence.[197] The principle of harmonious construction was described by the High Court of Australia in *Project Blue Sky Inc* v. *Australian Broadcasting Authority*:

> A legislative instrument must be construed on the prima facie basis that its provisions are intended to give effect to harmonious goals. Where conflict appears to arise from the language of particular provisions, the conflict must be alleviated, so far as possible, by adjusting the meaning of the competing provisions to achieve that result which will best give effect to the purposes and language of those provisions while maintaining the unity of all the statutory provisions.[198]

If a harmonious construction cannot be achieved and the conflicting objects of the legislation will materially affect the interpretation of an ambiguous provision, the court must seek to determine which of the objects is pre-eminent.[199] This is achieved through an application of the ordinary principles of statutory interpretation within the relevant jurisdiction, but will ordinarily require an analysis of the text, context of the Act as a whole (including any relevant extrinsic materials) and the purpose of the legislation. Where the dominant purpose is still not apparent, it is likely that the court will use a similar process to the interpretation of a company memorandum: that is, the earliest objectives in the section will likely be considered to be the dominant purpose, with the later objectives being considered ancillary.[200] Similar processes are used in both civil law and mixed law jurisdictions. All of this could be avoided, however, if this issue were given more consideration or more direction were provided on the relative prioritisation of objectives.

4.6 CONCLUSION

This chapter has highlighted that, beyond the narrow economic justifications for intervening in the renewable energy sector based on market failures, countries are seeking to achieve a very broad range of legislative objectives through their renewable energy laws. In contrast to the previous research

[197] Ruth Sullivan, *Sullivan and Driedger on the Construction of Statutes* (4th edn, Butterworths, 2002) 262 quoted in Mark Leeming, *Resolving Conflicts of Laws* (Federation Press, 2011) 47.

[198] (1998) 194 CLR 355, [70] (McHugh, Gummow, Kirby and Hayne JJ). Similar principles exist in other common law jurisdictions: *Saulnier v. Royal Bank of Canada WBLI Inc* [2008] 3 SCR 166, [16] (Binnie J, McLachlin CJ and LeBel, Deschamps, Fish, Abella, Charron and Rothstein JJ concurring); *Attorney-General* v. *Sillem* (1864) 159 ER 178, 217 (Pollock LCB).

[199] *Institute of Patent Agents* v. *Lockwood* [1894] AC 347, 360 (Herschell LJ).

[200] *Re Haven Gold Mining Co* (1882) 20 Ch D 151.

suggesting a high degree of commonality in the legislative objectives adopted around the world, some 28 categories of legislative objective were identified in the laws of the 113 countries with renewable energy laws. Further, the fact that a theme of legislative objective sought to address one of the sources of market failure identified in the previous chapter did not mean that it necessarily was prioritised over those themes that merely sought to address domestic market barriers or other goals. The exceptions to this rule are the security themed objectives, which were often prioritised over the objectives in the other themes. However, the education, training and research themed objectives received a higher weighted rank (i.e. lower priority) and were cited less frequently than any of the sectoral themed objectives. This process was even evident in the environmental themed objectives. Indeed, the objective that most closely addressed the unpriced environmental externalities within the energy sector, 'reduce greenhouse gas emissions and address climate change', was less likely to be cited than 'environmental protection' and had a higher weighted rank of 5.89 (i.e. lower priority).

This chapter showed that there is significant national variance in the legislative objectives of different countries in their renewable energy laws, which has an impact upon their scope and implementation. The sheer number of legislative objectives that countries are seeking to achieve also means that it is likely that any normative power of the objectives will be diminished and that conflicts may arise between them. This should prompt two questions. First, are countries that are seeking to achieve multiple and conflicting legislative objectives in their national renewable energy laws far beyond the narrow economic justification over-reaching in their regulation? Certainly, the economists would argue that this is the case and, if they are correct, this may lessen the efficacy and legitimacy of these laws. Second, does the extent of the variance within the legislative objectives adopted by different countries provide an explanation for why countries have thus far been unwilling to engage in legal harmonisation or convergence of their renewable energy laws despite the globalisation of renewable energy technologies? This suggests that there are still strong domestic factors at play within the renewable energy sector, which closely reflect a country's natural resource endowment, economy, political and market structure, institutions and culture. Given the important role that energy plays in everyday life and the sensitivity of the electorate to energy issues, this is an area of their legislative competence that many countries are reluctant to give up.

The countries with the greatest degree of similarity in their legislative objectives were the EU Member States. This reflects an element of normative convergence (that is, an alignment of cultural norms) derived from a regional approach to the accelerated deployment of renewable energy. For the other countries where there was clear normative divergence, the differences in the legislative objectives, when coupled with the use of different regulatory support mechanisms, may have a profound impact on the scope, implementation and outcomes of these laws.

The other problem that becomes apparent from a study of the legislative objectives is the effect that uncertainty might have on the sector. Some uncertainty within the sector is inevitable. For example, the future cost of renewable energy generation and technologies, the future cost of fossil fuel generation, the degree and location of economic benefits and the cost of climate change and other environmental and geopolitical risks that renewable energy laws are seeking to mitigate are not known with certainty.[201] This is no different from legislating in other areas. However, the failure to clearly articulate and appropriately prioritise the legislative objectives risks adding a further layer of uncertainty that could be avoided.

Countries need to consider how many legislative objectives they are trying to achieve, how they will prioritise them and how they will interact. They also need to consider whether renewable energy law is the most efficient and effective way of achieving those goals. Courts will not have to resort to the principles of statutory interpretation to resolve the priorities of a piece of legislation if it is well drafted. Countries also need to view the legislative objectives in renewable energy legislation as long-term goals.[202] In this way, the objectives will be less likely to be subject to frequent amendment and will send clear and stable signals to the market about what the country is attempting to achieve through its renewable energy law.

201 Grace et al., above n 36, 30.
202 Ibid 32.

PART III

WHAT ROLE DO REGULATORY SUPPORT MECHANISMS PLAY IN NATIONAL RENEWABLE ENERGY LAWS? A CASE OF SUBSTANTIVE DIVERGENCE

5

How Do Countries Regulate to Support Renewable Energy?

As with both the definition of renewable energy and the legislative objectives adopted, there is also considerable variation in the regulatory support mechanisms contained within the national renewable energy laws of different countries. This reflects the variety of domestic market barriers that exist within different countries (and even in some instances, regions) and the broad range of objectives that governments are seeking to achieve through these mechanisms.

5.1 THE SELECTION OF REGULATORY SUPPORT MECHANISMS

A number of factors may be relevant when considering which regulatory support mechanisms should be adopted within a particular country, for example:

1. Is the regulatory support mechanism designed to target the price or quantity of renewable energy to be deployed?[1]
2. Is the regulatory support mechanism designed to target the supply side or demand side of the renewable energy market?[2]

[1] Philippe Menanteau, Dominique Finon and Marie-Laure Lamy, 'Prices Versus Quantities: Choosing Policies for Promoting the Development of Renewable Energy' (2003) 31 *Energy Policy* 799; Reinhard Haas et al., 'A Historical Review of Promotion Strategies for Electricity from Renewable Energy Sources in EU Countries' (2011) 15 *Renewable and Sustainable Energy Reviews* 1003, 1011.

[2] Lincoln L Davies, 'Reconciling Renewable Portfolio Standards and Feed-in Tariffs' (2012) 32 *Utah Environmental Law Review* 311, 319–20; Richard L Ottinger, Lily Matthews and Nadia Elizabeth Czachor, 'Renewable Energy National Legislation: Challenges and Opportunities' in Donald N Zillman et al. (eds.), *Beyond the Carbon Economy: Energy Law in Transition* (Oxford University Press, 2008) 183, 192–200.

3. Is the regulatory support mechanism going to be compulsory or a voluntary approach?[3]
4. Is the regulatory support mechanism going to attempt to 'pick winners' or be technology neutral?[4]
5. Is the regulatory support mechanism going to be available on an industry-wide basis or will it target projects of a particular size or type?[5]
6. Is the regulatory support mechanism going to be capped by MW or GW *installed* (i.e. capacity), MW or GW *generated* or a *fixed budgetary pool* or some other means?[6]
7. If a price-based strategy is chosen, is the price to be fixed (e.g. such as a carbon price) subject to competitive bidding via an auction or endogenous (i.e. subject to market fluctuations due to trading of an instrument)?[7]

[3] See e.g. Robert C Grace, Deborah A Donovan and Leah L Melnick, *When Renewable Energy Policy Objectives Conflict: A Guide for Policymakers* (2011) National Regulatory Research Institute <http://nrri.org/download/2011–17-when-renewable-energy-policy-objectives-conflict/#>; Ottinger et al., above n 2, 193–9; Reinhard Haas et al., 'Promoting Electricity from Renewable Energy Sources – Lessons Learned from the EU, US and Japan' in Fereidoon P Siosanshi (ed.), *Competitive Electricity Markets: Design, Implementation, Performance* (Elsevier Science, 2008) 419, 425; Haas et al., 'A Historical Review of Promotion Strategies', above n 1, 1012; Trent Berry and Mark Jaccard, 'The Renewable Portfolio Standard: Design Considerations and an Implementation Survey' (2001) 29 *Energy Policy* 263, 264–5, 268; Simone Espey, 'Renewables Portfolio Standard: A Means for Trade with Electricity from Renewable Energy Sources?' (2001) 29 *Energy Policy* 557, 558; Adrian Bradbrook, 'Green Power Schemes: The Need for a Legislative Base' (2002) 26 *Melbourne University Law Review* 15, 20–30; Janet Sawin, 'National Policy Instruments: Policy Lessons for the Advancement & Diffusion of Renewable Energy Technologies Around the World' (Paper presented at the International Conference for Renewable Energies, Bonn, 2004) 2; Benjamin K Sovacool, *Renewable Electricity for Southeast Asia: Designing the Right Policy Architecture* (Lee Kuan Yew School of Public Policy, National University of Singapore, 2009) 16; Warren Leon and Clean Energy States Alliance, 'Designing the Right RPS: A Guide to Selecting Goals and Program Options for a Renewable Portfolio Standard' (Guide, State-Federal RPS Collective and the National Association of Regulatory Utility Commissioners, 2012).

[4] Catherine Mitchell, *Energy, Climate and Environment Series: The Political Economy of Sustainable Energy* (Palgrave Macmillan, 2010) 39–57.

[5] Pere Mir-Artigues and Pablo del Río, 'Combining Tariffs, Investment Subsidies and Soft Loans in a Renewable Electricity Deployment Policy' (2014) 69 *Energy Policy* 430.

[6] See e.g. New South Wales Auditor-General, *NSW Auditor-General's Special Report into the NSW Solar Bonus Scheme* (New South Wales Audit Office, 2011) 24.

[7] Carolyn Fischer and Louis Preonas, 'Combining Policies for Renewable Energy: Is the Whole Less than the Sum of Its Parts?' (2010) *International Review of Environmental and Resource Economics* 51, 69–70; Aviel Verbruggen and Volkmar Lauber, 'Assessing the Performance of Renewable Electricity Support Instruments' (2012) 45 *Energy Policy* 635, 642; Carolyn Fischer and Richard G Newell, 'Environmental and Technology Policies for Climate Mitigation' (2008) 55 *Journal of Environmental Economics and Management* 142, 150–1.

8. Should the cost of the regulatory support mechanism be borne by conventional utility companies, end-consumers or taxpayers more broadly?[8]
9. Is the country a member of a regional organisation such as the EU that may place restrictions such as the State Aid rules on the use of regulatory support mechanisms?[9]
10. Is the country a member of the World Trade Organization (WTO) so subject to the Agreement on Subsidies and Countervailing Measures?[10]
11. Is the country a member of the Energy Charter Treaty or another bilateral or multilateral agreement that may impact upon the design, implementation, or any subsequent amendment to, their national renewable energy law?

Once these questions have been answered, governments have a range of regulatory support mechanisms to select from, and then modify, as appropriate. A number of studies have sought to present classification systems to help foster an understanding of the regulatory support mechanisms used in renewable energy laws.

8 Verbruggen et al., above n 7, 641; Steffen Jenner et al., 'What Drives States to Support Renewable Energy?' (2012) 33(2) *Energy Journal* 1, 4; Reinhard Haas et al., 'How to Promote Renewable Energy Systems Successfully and Effectively' (2004) 32 *Energy Policy* 833, 839; Mitchell, *The Political Economy of Sustainable Energy*, above n 4, 183; Sawin, above n 3, 12–13.

9 *PreussenElektra AG v. Schhleswag AG* (C-379/98) [2001] EUECJ 160; European Commission, 'State Aid: Commission Opens In-depth Inquiry into Support for Energy-intensive Companies Benefitting from a Reduced Renewables Surcharge' (Press Release, Brussels, 18 December 2013); Dave Keating, *Commission Unveils Overhaul of Renewable Energy Subsidies* (9 April 2014) European Voice <www.europeanvoice.com/article/2014/april/commission-unveils-overhaul-of-renewable-energy-subsidies/80450.aspx>; Kim Talus, 'Treaty Law and the Energy Sector' in Kim Talus (ed.), *EU Energy Law and Policy: A Critical Account* (Oxford University Press, 2013) 110; Angus Johnston et al., 'Rethinking the Scope and Necessity of Energy Subsidies in the United Kingdom' (2014) 3 *Energy Research & Social Science* 1, 3.

10 *Marrakesh Agreement Establishing the World Trade Organisation*, opened for signature 15 April 1994, 1867 UNTS 3 (entered into force 1 January 1995) annex 1A ('Subsidies and Countervailing Measures'); Marie Wilke, 'Feed-in Tariffs for Renewable Energy and WTO Subsidy Rules: An Initial Legal Review' (Issue Paper No. 4, Institutional Centre for Trade and Sustainable Development, November 2011); Office of the United States Trade Representative: Executive Office of the President, 'China Ends Wind Power Equipment Subsidies Challenged by the United States in WTO Dispute' (Press Release, 6 June 2011) <https://ustr.gov/about-us/policy-offices/press-office/press-releases/2011/june/china-ends-wind-power-equipment-subsidies-challenged>; B. Olmos Giupponi, 'Mapping Emerging Countries' Role in Renewable Energy Trade Disputes' (2015) 12(3) *Transnational Dispute Management* 1; Timothy Meyer, 'Explaining Energy Disputes at the World Trade Organization' (2017) 17 International Environmental Agreements 391; Rafael Leal-Arcas and Andrew Filis, 'Renewable Energy Disputes in the World Trade Organization' (2015) 12(3) *Transnational Dispute Management* 1.

5.1.1 *Primary Instruments Versus Secondary Instruments*

The first basis for classification assesses the economic incentives contained in the instrument to determine whether a regulatory support mechanism is a primary or secondary instrument. Primary instruments are those that are generally national in their scope and applicable to all technologies (although the incentives may be banded in recognition of their degree of commercialisation). In contrast, secondary instruments are much more limited in their scope, with restrictions on the size of qualifying projects and the technologies which qualify for support. On this basis, the following regulatory support mechanisms are typically characterised as primary instruments: feed-in tariffs, quota obligations with tradeable green certificates (TGC) and competitive tendering; while secondary instruments include investment subsidies, fiscal incentives and soft loans.[11]

This distinction has been adopted widely within the existing body of literature, although often under different names, as was recognised by Mir-Artigues and del Rio:

> The distinction between primary and secondary instrument is a widespread and classical one in the RES-E support literature, although with different names, 'dominating instruments' in Ragwitz (2012), 'main support schemes' in Klessmann and Lovinfosse (2012), Teckenburg et al. (2012) and IEA/IRENA (2013) and 'primary' and 'secondary' instruments in Ragwitz et al. (2012), Huber et al. (2004) and Del Rio and Gual (2004).[12]

This classification system appears to be valid, as many of the primary instruments overlap in their coverage of legislative objectives so countries will tend to focus on only one or two primary instruments to avoid overregulation and overcompensation of market participants. In addition to the primary instruments adopted, most countries also use a number of secondary instruments to provide targeted support to particular technologies or smaller projects.

5.1.2 *Support for Investments Versus Operating Support*

An alternative basis for classification is propounded by Fräss-Ehrfeld, who suggests that government subsidies can be divided between those that support investment in renewable energy such as capital grants, tax exemptions or rebates on equipment purchases, and those that support the operation of

11 Mir-Artigues et al., above n 5.

12 Ibid fn 2.

renewable energy projects.[13] Fräss-Ehrfeld states that the regulatory support mechanisms that fall within this latter category include 'price subsidies, green certificates, tender schemes, and tax exemptions or reductions on the production of electricity'.[14] This classification does not seem to add a lot of value to countries selecting between different regulatory support mechanisms, as arguably the ultimate aim of all regulatory support mechanisms is to accelerate the deployment of renewable energy projects that generate electricity. In distinguishing between investment support and operating support, this classification may not be sufficiently refined to differentiate between the support of projects that generate electricity, as opposed to simply installed capacity that may not be connected to the grid.

This issue of grid-connected projects versus installed capacity is a particular problem in some countries such as China. For example, in 2009 when the Chinese wind feed-in tariff was introduced, China had not constructed sufficient high voltage transmission lines to transport this power without significant load losses from the west of China, where it was generated, to the east of China, where there was substantial demand.[15] Further, many areas where wind farms were constructed could not immediately be connected to the transmission grid without either the grid being upgraded or new transmission and distribution lines being installed.[16] This meant that a number of newly constructed wind farms became, at least temporarily, 'stranded assets' unable to generate electricity because it could not be transported to consumers. Indeed, in 2011 it was estimated that nearly 30 per cent of Chinese wind farms were not connected to the grid due to a lack of additional capacity on the transmission and distribution networks or the problems with frequency management which occur when rapidly bringing online significant intermittent generation.[17] This issue of countries installing additional capacity that is neither distributed generation nor grid-connected has been so problematic that in the past it prompted REN-21

[13] Clarisse Fräss-Ehrfeld, *Renewable Energy Sources: A Chance to Combat Climate Change* (Wolters Kluwer, 2009) 262–3.

[14] Ibid 263.

[15] Anthony Kim and Olio Wang, 'Sinovel Legal Action Casts Shadow over Potential Export Opportunities in China's Wind Power Industry', *Financial Times* (online), 19 September 2011 <www.ft.com/cms/s/2/83fef47e-e2cc-11e0-93d9-00144feabdc0.html#ixzz1dy6b1n00>.

[16] Kat Cheung, *Integration of Renewables - Status and Challenges in China* (Working Paper, IEA/OECD, 2011).

[17] Robert Crowe, 'China Calls on A123 to Aid Wind Integration', *Renewable Energy World* (online), 12 August 2011 <www.renewableenergyworld.com/rea/news/article/2011/08/china-calls-on-a123-to-aid-wind-integration>.

to alter the way that they calculated and reported renewable energy statistics.[18] As a result, the use of this basis of classification without the addition of further distinctions or refinements seems flawed.

5.1.3 *Supply Side Strategies Versus Demand Side Strategies*

A third basis for classifying the existing regulatory support mechanisms is whether they target the supply side or the demand side of the renewable energy market. Davies has stated that supply-side or market 'push' mechanisms 'seek to promote the quantity of a given type of technology . . . to augment the amount of a resource or a technology that is available for commercial use'.[19] Examples of supply-side regulatory support mechanisms include: '(a) conducting basic applied research and development on energy technologies; (b) building large test or prototype facilities; (c) having the government procure large amounts of an experimental technology; and (d) investor tax credits that spur innovation on a given technology'.[20] In contrast, demand-side or market 'pull' mechanisms seek to foster increased demand for renewable energy technologies, which in turn should lead to more technologies coming to the market to meet that demand. Examples of demand-side regulatory support mechanisms include: '(a) creating markets for [renewable energy – sic] through production tax credits; (b) establishing rate-based or purchase-based incentives such as higher rates of return or tariffs; (c) promoting technologies through training or information and awareness campaigns'.[21]

There is some disagreement between scholars as to whether two of the most commonly adopted regulatory support mechanisms, renewable portfolio standards/quota obligations/green certificate trading and feed-in tariffs, are in fact both demand-side strategies, or whether feed-in tariffs should be conceived of as a supply-side strategy.[22] The Energy and Resources Institute

18 REN21 Secretariat, 'Renewables 2013 Global Status Report' (Report, Renewable Energy Policy Network for the 21st Century, 2013) 126–7.

19 Davies, 'Reconciling Renewable Portfolio Standards and Feed-in Tariffs', above n 2, 319.

20 Sovacool, *Renewable Electricity for Southeast Asia*, above n 3, 13.

21 Ibid.

22 See e.g. Davies who advocates that both quantity-driven policies such as renewable portfolio standards and price-driven strategies such as feed-in tariffs are demand-side or market 'pull' mechanisms: Davies, 'Reconciling Renewable Portfolio Standards and Feed-in Tariffs', above n 2, 320. In contrast, The Energy and Resources Institute advocates that feed-in tariffs should be conceptualised as a supply-side mechanism, with quota obligations and renewable portfolio standards with green tradeable certificates thought of as demand-side mechanisms: The Energy and Resources Institute, 'International Policy and Regulatory Regimes for Promoting Renewable Energy Based Electricity Generation' in The Energy and Resources Institute (TERI) (ed.), *Policy & Regulatory Approaches for Promoting RE Power* (REEEP, Unknown) 4.

distinguishes the most common mechanisms on this basis and also on whether they are generation based (that is, focus on kWh or MWh) or capacity based (that is, focus on kW or MW installed).[23] By classifying regulatory support mechanisms as targeting the supply side or demand side and focusing on whether the investment is focused on building generation or capacity, this approach overcomes the issues with Fräss-Ehrfeld's classification. Regardless of the characterisation of renewable portfolio standards/quota obligations/green certificate trading and feed-in tariffs, the classification of regulatory support mechanisms as supply-side and demand-side mechanisms is arguably a helpful one. This classification is one that is commonly used by countries when devising regulatory and policy strategies for accelerating technological development and innovation so there should be a degree of familiarity with this approach.

5.1.4 *Price-Driven Strategies Versus Quantity-Driven Strategies*

A fourth basis for classifying the existing regulatory support mechanisms is whether they are price-driven or quantity-driven. Once this initial distinction has been made, further distinctions are often made between regulatory and voluntary mechanisms, investment or generation focused, and direct and indirect mechanisms. These distinctions have also received significant support from eminent scholars working in the field such as Menanteau, Finon and Lamy,[24] Verbruggen and Lauber,[25] Haas et al. (2008)[26] and Haas et al. (2011).[27]

Haas et al. have drawn upon the work of Menanteau, Finon and Lamy[28] to classify the commonly adopted primary and secondary regulatory support mechanisms using these distinctions, as shown in Table 5.1. This approach recognises that there are fundamentally four different ways of promoting electricity derived from renewable energy sources:

1. regulatory price-driven mechanisms;
2. regulatory quantity-driven mechanisms;
3. voluntary mechanisms; and
4. indirect mechanisms.

23 See, TERI (ed.), above n 22, 4.
24 Menanteau et al., above n 1.
25 Verbruggen et al., above n 7, 637–8.
26 Haas et al., 'Promoting Electricity from Renewable Energy Sources', above n 3, 424–5.
27 Haas et al., 'A Historical Review of Promotion Strategies', above n 1, 1011–2.
28 Menanteau et al., above n 1.

TABLE 5.1 *Fundamental Types of Promotion Strategies*[29]

		Direct Price-driven	Direct Quantity-driven	Indirect
Regulatory	Investment focused	Investment incentives Tax credits Low interest/soft loans	Tendering system for investment grant	Environmental taxes Simplification of authorisation procedures Connection charges, balancing costs
Regulatory	Generation based	(Fixed) Feed-in tariffs Fixed premium system	Tendering system for long-term contracts Tradable green certificate system	
Voluntary	Investment focused	Shareholder programmes Contribution programmes		
Voluntary	Generation based	Green tariffs		Voluntary agreements

5.1.4.1 Regulatory Price-Driven Mechanisms

The most commonly adopted regulatory support mechanism is the feed-in tariff, which is an example of a regulatory price-driven mechanism. Price-driven mechanisms do not set quantity goals or targets or quotas for the amount of renewable energy to be generated. Instead, these mechanisms provide renewable energy generators with either a fixed amount per kW of capacity installed or kWh generated or a premium on top of the electricity price for each kWh generated.[30]

5.1.4.2 Regulatory Quantity-Driven Mechanisms

The main alternative to price-driven mechanisms for primary instruments is quantity-driven mechanisms. Quantity-driven mechanisms such as renewable portfolio standards, quota obligations and renewable energy targets work by establishing a quota (either in terms of MW or GW of renewable electricity to

[29] Haas et al., 'A historical review of promotion strategies', above n 1, 1011–2.

[30] See e.g. Menanteau et al., above n 1; Haas et al., 'Promoting electricity from renewable energy sources', above n 3, 424–5.

be generated or the percentage share of the electricity generation to be derived from renewable energy sources). Most quotas have a specific date by which the quota has to be achieved to ensure that electricity generated from renewable energy sources (or even following recent adaptions to quantity-driven mechanisms, specific emerging renewable energy technologies) achieves the desired amount of market penetration.[31]

The argument about whether regulatory price-driven or quantity-driven mechanisms are more effective has been the source of much academic debate and will be discussed further below.

5.1.4.3 Voluntary Approaches

Not all support mechanisms are regulated, with a number of voluntary approaches also used to promote the accelerated deployment of renewable energy. Voluntary approaches are predicated on the willingness of investors either to make an upfront capital investment into the renewable energy project[32] or, alternatively, to pay a volumetric premium for electricity generated from renewable energy sources. The latter mechanism is often called a 'green power scheme', 'green marketing' or a 'green tariff' in the literature.[33]

5.1.4.4 Indirect Mechanisms

The last distinction made under this system of classification is whether the mechanism is directly focused on the immediate stimulation of electricity generated from renewable energy sources or whether it is a more indirect mechanism. Indirect mechanisms are designed to make the market environment for renewable energy more attractive in the long term. Indirect mechanisms include strategies such as providing preferential permitting and siting, preferential grid connection, fast-tracked administrative approvals, the imposition of taxes or levies on brown electricity such as carbon taxes, sulphur taxes or pollution taxes and the removal of subsidies on conventional fossil fuels. These indirect mechanisms all aim to reduce the market barriers to new entry and transaction costs for market participants.[34]

31 Ibid.

32 Haas et al., 'Promoting Electricity from Renewable Energy Sources', above n 3, 425.

33 Bradbrook, above n 3; Nico H Van der Linden et al., *Review of International Experience with Renewable Energy Obligation Support Mechanisms* (Dutch Ministry of Economic Affairs, 2005) 12; Haas et al., 'Promoting Electricity from Renewable Energy Sources', above n 3, 425.

34 Haas et al., 'Promoting Electricity from Renewable Energy Sources', above n 3, 425–6.

5.2 TYPES OF REGULATORY SUPPORT MECHANISMS USED IN THE RENEWABLE ENERGY SECTOR

Once countries have decided the basic design features of their regulatory support mechanisms to encourage the accelerated deployment of electricity generated from renewable sources, they then have a number of common mechanisms from which to choose. These regulatory support mechanisms include feed-in tariffs, feed-in premiums, renewable portfolio standards/quota obligations, green certificate trading/renewable energy credits, competitive tendering (auction bidding), net metering, subsidies, loans, rebates, investment tax credits, production tax credits, green power schemes, grants, research and development support and other indirect mechanisms.

One of the difficulties in comparing these support mechanisms is that while each type of mechanism possesses some common characteristics, their design and implementation differs in every country. This is the case even within the EU, with the Member States able to decide on the design and implementation of their own regulatory support mechanisms within the framework of the EU Renewable Energy Directive.[35] There are a number of reasons for this, including the differences in the indigenous renewable and non-renewable energy sources in each country, the structure of the national energy markets, and the varying legislative objectives adopted in the national renewable energy laws. A further complicating factor is that because most countries use several regulatory support mechanisms in combination, the mechanisms and other policies interact with each other, making it difficult to isolate the precise impacts of each mechanism. For this reason, the common features of each regulatory support mechanism will be discussed below, before a discussion of the means of evaluating the relative success of the mechanisms and their use in combination.

5.2.1 *Feed-in Tariffs*

Once the most commonly adopted regulatory support mechanism, feed-in tariffs (FITs)[36] have recently been surpassed by competitive tendering as the

35 Directive 2009/28/EC of the European Parliament and of the Council of 23 April 2009 on the promotion of the use of energy from renewable sources and amending and subsequently repealing Directives 2001/77/EC and 2003/30/EC (Text with EEA relevance) [2009] OJ L 140/16.

36 'Feed-in tariffs (FITs) are also known as Standard Offer Contracts, Feed Laws, Minimum Price Payments, Renewable Energy Payments, and Advanced Renewable Tariffs': Toby Couture and Yves Gagnon, 'An Analysis of Feed-in Tariff Remuneration Models: Implications for Renewable Energy Investment' (2010) 38 *Energy Policy* 955.

tool of choice to support the accelerated deployment of renewable energy.[37] FITs are price-based mechanisms, which in their most basic form offer fixed preferential prices for each kWh (or MWh) generated from renewable energy sources (a 'gross FIT') or for each kWh of renewable electricity transmitted to the grid (a 'net FIT').[38] Historically, they were often coupled with priority dispatch to the grid and a binding purchase obligation upon electricity supply companies,[39] though these conditions are now less common in modern FITs. Where conditions are available, they are normally contractually guaranteed for a minimum period, which may range from eight to twenty years to provide long-term security for investors and market stability.[40] FIT prices are normally reviewed on an annual or bi-annual basis, with project developers benefiting from the rates assigned to that year's vintage for the life of the feed-in tariff contract.[41] This process is also sometimes referred to as 'grandfathering'. Grandfathering provides long-term security and stability in the market, as it ensures that renewable energy projects continue to be awarded the FIT rates of their vintage, even if the FIT rate for subsequent vintages has been reduced to reflect reductions in equipment and capital costs, or the benefits of new innovations and learning.[42] Some countries such as Austria use budgetary caps to control the costs of their FIT, with other countries such as Cyprus and Portugal using caps on the amount of production as a means of exercising

[37] REN21 Secretariat, 'Renewables 2018 Global Status Report' (Report, Renewable Energy Policy Network for the 21st Century, 2018) 64–7.

[38] For a greater discussion of gross and net FITs, see Parliamentary Library of the Commonwealth of Australia, *Feed-in Tariffs* (21 December 2011) Parliament of Australia <www.aph.gov.au/About_Parliament/Parliamentary_Departments/Parliamentary_Library/Browse_by_Topic/ClimateChangeold/governance/domestic/national/feed >.

[39] Ottinger et al., above n 2, 192; Van der Linden et al., above n 34, 11; Lincoln L Davies, 'Incentivizing Renewable Energy Deployment: Renewable Portfolio Standards and Feed-in Tariffs' (2011) 1 *KLRI Journal of Law and Legislation* 39, 54. Though note that Becker et al. do not believe this is a necessary condition for a FIT: Bastian Becker and Doris Fischer, 'Promoting Renewable Electricity Generation in Emerging Economies' (2013) 56 *Energy Policy* 446, 447.

[40] Arne Klein et al., *Evaluation of Different Feed-in Tariff Design Options – Best Practice Paper for the International Feed-in Cooperation* (Ministry for the Environment, Nature Conservation and Nuclear Safety, 2010); Verbruggen et al., above n 7, 637; Frässe-Ehrfeld, above n 13, 264; Kate Loynes, *Overview of Feed in Tariffs: A Quick Guide* (1 April 2014) Parliament of Australia <www.aph.gov.au/About_Parliament/Parliamentary_Departments/Parliamentary_Library/pubs/rp/rp1314/QG/Tariffs>; Judith Lipp, 'Lessons for Effective Renewable Electricity Policy from Denmark, Germany and the United Kingdom' (2007) 35 *Energy Policy* 5481, 5482.

[41] Verbruggen et al., above n 7, 637.

[42] Miguel Mendonça, 'FIT for Purpose: 21st Century Policy' (2007) 8(4) *Renewable Energy Focus* 60, 61.

budgetary restraint.[43] Budgetary control over feed-in tariffs remains a key issue, with a number of countries including China, Spain, Italy, Bulgaria and the Czech Republic recently having to either reduce their tariffs or close schemes early due to budget overrun.

The design of FITs has changed over time and varies considerably around the world. Initially, FITs were set at a flat rate, which did not discriminate between different technologies, in order to provide deployment at least cost. However, many FITs have now been modified to better encourage innovation and support emerging technologies. There are a number of different ways in which this can be done, including by providing differentiated or stepped tariffs for:

1. different sources of renewable energy in order to achieve diversity of supply;
2. different renewable energy technologies in order to ensure that there is adequate demand for promising technologies that have not yet been commercialised or reached widespread deployment.[44] This practice operates in countries such as Germany, Malaysia, Slovakia and Turkey, where the technologies are banded according to the relative degree of their commercialisation, with the FITs for each technology band reduced each year to account for improvements in the relative maturity of the technology.[45] The banding of technologies was implemented in order to ensure that the most commercialised technologies did not receive windfall profits under the FIT and that a diverse range of

43 RES Legal, *Compare Support Schemes* (2018) <www.res-legal.eu/compare-support-schemes/>; Lena Kitzing, Catherine Mitchell and Poul Erik Morthorst, 'Renewable Energy Policies in Europe: Converging or Diverging?' (2012) 51 *Energy Policy* 192, 194.

44 The higher tariffs on offer to less developed technologies reflect the higher costs associated with establishing one of these projects both in terms of initial capital costs for equipment and installation but also the higher capital cost of borrowing. See also Fischer and Preonas, above n 7, 58.

45 *See e.g. Gesetz für den Ausbau erneuerbarer Energien (Erneuerbare-Energien-Gesetz – EEG 2017)* [Renewable Energy Sources Act 2017] [German Federal Ministry for Economic Affairs and Energy translation from German], §§40–53; Sustainable Energy Development Authority Malaysia, *Feed-in Tariff Dashboard* (2018) <http://seda.gov.my/>; Vyhláška N°18/2017 Úradu pre reguláciu sieťových odvetví z 8. februára 2017, ktorou sa ustanovuje cenová regulácia v elektroenergetike a niektoré podmienky vykonávania regulovaných činností v elektroenergetike [Decree N° 18/2017 of the Regulatory Office for Network Industries dated 8 February 2017 establishing price regulation in the electricity sector and certain conditions for the implementation of regulated activities in the electricity sector], §10 (Slovakia); Council of Ministers Decision No. 2013/5625 published in the Official Gazette No. 28842, dated 5 December 2013 (Turkey).

renewable energy sources and technologies were supported.[46] This reflects the fact that the primary objective for many countries regulating to accelerate the deployment of renewable energy is to improve their energy security and diversity of supply;

3. the size of the installation to provide a means of supporting both small-scale and large-scale renewable energy projects. This can also be an important way of improving competition within the electricity market, removing market barriers and preventing the large-scale incumbents from dominating the emerging market;
4. the quality of the resource. There are historical examples of some countries such as Germany and Switzerland offering higher tariffs to projects located in areas with poorer renewable energy resources.[47] This practice continues to occur in China, with the National Energy Administration dividing the country into three regions based on their suitability for solar generation, which are then offered differential regional tariffs.[48] The justification for providing additional support to these projects is that these projects are less efficient and therefore will be viewed by commercial lenders as riskier, leading to higher interest rates charged on project loans and longer periods of time before the projects become profitable. This approach also means that regulators can set the average level of the feed-in price at a lower level, without the risk of overcompensating projects located at the sites with the highest resource quality or undercompensating project developers at sites with poorer resource quality. It can also help to address issues of grid constraint and connection issues by directing deployment to unconstrained regions. Despite this, this approach is likely to drive up the total costs of deploying renewable energy, while actively eschewing the basic principles of comparative advantage in site selection. These factors actively distort the renewable energy market in a manner contrary to basic economic principles. These costs need to be weighed up against the benefits that a tariff differentiated upon the quality of the resource may provide; and/or

46 IRENA, *Evaluating Policies in Support of the Deployment of Renewable Power* (IRENA, 2012) 10.

47 Couture et al., above n 37, 955.

48 Liu Bin, *China Solar Industry Struggles Through Sudden Subsidy Cuts* (15 August 2018) Climate Home News <www.climatechangenews.com/2018/08/15/china-solar-industry-struggles-sudden-subsidy-cuts/>.

5. the location of the project. This can encourage projects to be located in a diverse range of areas in order to achieve a good geographical spread and encourage flexibility of siting to improve social acceptance, to improve energy security or for political reasons.[49]

In 2010, Couture and Gagnon conducted a comprehensive study of the different remuneration models used by countries with feed-in tariffs, and the implications of these remuneration models on investment in the renewable energy sector. Their study found that feed-in tariffs, which operate on a market-independent basis, are much more common than feed-in premiums, which are market-dependent.[50] They also identified four basic models used for feed-in tariffs around the world: the fixed price model, the fixed price model with full or partial inflation adjustment, the front-end loaded model and the spot market gap model.[51]

5.2.1.1 Fixed Price Model

The fixed price model establishes a fixed, minimum price, which is paid for each kWh of electricity generated from renewable energy sources over the contract period. The factors considered in determining the fixed feed-in price generally include: the maturity of the technology, the cost of equipment and installation, the cost of capital to finance the project, licensing fees, the costs of operation and maintenance, the costs of feedstocks (relevant in the case of biomass and biogas) and a fair rate of return for investors (profitability).[52] This approach has been adopted in the Japanese FIT. An alternative way of setting the fixed feed-in tariff price is to use a 'value-based' assessment as a means of internalising the value of the positive externalities associated with electricity generation from renewable energy sources. The use of this method aims to 'securitise long-term grid, public health, and environmental benefits that clean distributed generation provide to a specific geographic area and/or location on the grid'.[53] The 'value-based' assessment method has not been widely adopted, with most countries preferring to price their feed-in tariff using the 'cost' method.

49 Ibid; Verbruggen et al., above n 7, 637.

50 Couture et al., above n 37, 956.

51 Ibid.

52 Rolf Wüstenhagen and Emanuel Menichetti, 'Strategic Choices for Renewable Energy Investment: Conceptual Framework and Opportunities for Further Research' (2012) 40 *Energy Policy* 1, 6; Frässe-Ehrfeld, above n 13, 268–9.

53 Pierre Bull, Noah Long and Cai Steger, 'Designing Feed-in Tariff Policies to Scale Clean Distributed Generation in the US' (2011) 24(3) *The Electricity Journal* 52, 53.

The fixed feed-in price will be retained for the length of the contract. Under this model, the fixed feed-in price is quarantined to isolate it from other market variables such as the retail price of electricity, inflation, movements in the consumer price index and the price of feedstock of fossil fuels.[54] Indeed, the only variation in the feed-in price available under this model is that some countries provide higher tariffs to renewable energy sources/technologies that are less developed, to ensure that their fixed feed-in tariff adequately compensates project developers for their higher project costs. The design of this model of feed-in tariff encourages project developers to deploy their projects early in the contract period.[55] This is because the feed-in price is frozen, meaning that over the life of the contract its value in real terms declines relative to the project costs. The benefits of the fixed feed-in price model are that it provides investors with certainty and makes accessing finance easier as the feed-in tariffs to be paid over the life of the project are known in advance. It is also comparatively simple to administer.[56]

5.2.1.2 Inflation Adjusted Fixed Price Model

The second remuneration model used in the feed-in tariffs is a variant on the fixed feed-in price model, which either fully or partially adjusts for inflation. The design of this model, which is used in countries such as the Czech Republic and Hungary, seeks to overcome the decline in real terms found in the basic fixed feed-in price model. This means that unlike the basic fixed feed-in price model, the inflation adjusted fixed price model is likely to provide higher levels of remuneration near the end of a project's life.[57] For most projects, this is the period in which their capital equipment and installation costs have been paid and thus, much of the revenue generated will be profit. Couture and Gagnon have argued that 'this puts an undue burden on the electricity ratepayer in the long-term, by requiring continually high payments until the end of the contract term'.[58] Despite this, the inflation adjusted feed-in price model presents three advantages to regulators. First, by linking the feed-in price to inflation, it removes some of the market risk and provides additional security to investors, making it easier overall to obtain project finance. Second, due to the fact that a lower feed-in price is paid at the beginning of the term of the contract, this model may increase the levels of social acceptance and be easier to implement politically. Third, the market

54 Couture et al., above n 37, 956.
55 Ibid 956–7.
56 Menanteau et al., above n 1, 807.
57 Couture et al., above n 37, 957.
58 Ibid.

standard for Power Purchase Agreements (PPAs) is that they will be indexed to the CPI. As a result, it is arguable that the fixed feed-in price model adjusted for inflation reflects market-standard contractual terms within the electricity industry.[59]

5.2.1.3 Front-End Loaded Model

The third model of feed-in tariff is the front-end loaded model. In this model, the feed-in price is subject to high prices at the start of the contract, which then decrease over the life of the contract.[60] An example of this model is found in the Belarusian FIT, which has a high initial FIT multiplier for the first ten years of the project, before stepping down the multiplier at the ten-year anniversary date and then again at the twenty-year anniversary date for the remaining life of the contract.[61] The outcome of this model is similar to that of the basic fixed feed-in price model, but the effect may be more pronounced where the price decreases exceed the level of inflation. This model is beneficial as it provides greater levels of funding during the initial and highly capital-intensive phases of the project, enabling project developers to repay their project loans over a shorter period. The use of this model for differentiated tariffs has the benefit of encouraging innovation and improvements in production performance projects in lower resource quality areas to make up for the declining tariff levels. However, the high upfront cost burden may make the introduction of a front-end loaded model politically more difficult than the other FIT models presented here.

5.2.1.4 Spot Market Gap Model

The fourth feed-in tariff model is the spot market gap model, which is also sometimes referred to as the 'target price feed-in tariff'[62] model. The first step under this model is that the government must decide upon the desired level of generation from that technology class. This is then used to determine the required feed-in price for each class of technology in order to achieve the desired level of generation (the 'target price'). This determination is normally made in accordance with the principles of the national renewable energy

59 Ibid 958.

60 Haas et al., 'A Historical Review of Promotion Strategies', above n 1, 1014.

61 Resolution of the Council of Ministers of the Republic of Belarus of 07.08.2015 №45 'On the tariffs for electricity produced from renewable energy sources on the territory of the Republic of Belarus by individual entrepreneurs and legal entities who are not members of the State Electricity Production Association "Belenergo", and released to supply companies of this association'.

62 Kitzing et al., above n 44, 194.

strategy. Once the target price has been determined, renewable electricity generators have to market their own power on the spot market (though in many cases, this model retains a purchase obligation). The feed-in tariff payment is calculated by subtracting the spot market price from the target price, with the government making up the shortfall from consolidated revenue.[63] As a result, this model has the benefit of shifting the cost of renewable energy deployment not to end-use electricity consumers like many models, but ultimately to the taxpayer on the basis of their income.[64] This means that unlike many other regulatory support mechanisms used to support the accelerated deployment of renewable energy, this is not a regressive model and does not directly impact electricity prices. This model does however create a disconnect between those who pay for the feed-in tariff and those who benefit from its existence. This is because the most energy intensive industries, that are arguably receiving the greatest benefit from increased deployment of renewable energy through improved sector performance and energy security, do not necessarily bear the largest financial burden. It also arguably increases the risk for renewable energy project developers. This is because the spot market gap model is 'contingent on a specific budgetary allocation ... which may be exhausted, or fail to be renewed, by the time a proposed project begins supplying electricity to the grid'.[65] A further challenge in this model is a requirement that electricity generators market their own power on the spot market. This requirement encourages the market integration of electricity generated from renewable energy sources into the conventional electricity market. However, the transaction costs and administrative burden associated with participating in the spot market may exclude smaller-scale renewable energy generators from the market.[66]

In addition to the four basic models of feed-in tariffs detailed above, there have been numerous variations proposed and/or implemented in a number of countries to reflect their national priorities and renewable energy resource. For example in Latvia, prior to the suspension of its FIT in 2016 due to corruption concerns, the feed-in tariffs were linked to specific market indicators including the price of natural gas, and prior to 1 January 2014, the exchange rate of the Lats to the Euro.[67] Meanwhile other countries such as Hungary and the Ukraine have differentiated tariffs for their renewable energy

63 Couture et al., above n 37, 959.
64 See Menanteau and Sawin for alternative ways of balancing the costs of a FIT between electricity consumers and taxpayers: Menanteau et al., above n 1, 802; Sawin, above n 3, 5.
65 Couture et al., above n 37, 959.
66 Ibid.
67 Kitzing et al., above n 44, 194.

based on the time of day and whether the electricity is being consumed during a peak or off-peak period.[68] In addition, Lesser et al. have proposed the design of a 'two-part FIT', consisting of a capacity payment determined by an auction process and an energy payment that is linked to the spot market price of electricity.[69] The 'two-part FIT' is yet to be adopted by any country, but it exemplifies the innovative structuring of regulatory support mechanisms within the sector.

5.2.1.5 The Advantages and Disadvantages of Using Feed-in Tariffs

Feed-in tariffs were historically considered one of the most effective regulatory support mechanisms due to their ability to rapidly accelerate deployment of renewable energy generation in a cost-effective manner.[70] They offer a number of advantages including that they do not require government financial support,[71] they may be structured for large and small generators, different technologies and renewable energy sources through banding, and if the rates decline over time, they may be effective at putting pressure to lower costs.[72] A further advantage is that FITs are usually simple to administer and have comparatively low transaction costs.[73] However, their success depends on the tariffs being set at a sufficiently high level to encourage investment,[74] a stable and predictable regulatory environment for the FITs[75] and the electricity generated being able to access the transmission and distribution networks. FITs do present a number of disadvantages including the risk that the tariff will not be set at the correct level due to the difficulties associated with getting up-to-date information on generation costs using different energy

68 RES Legal, *Compare Support Schemes* (2018) <www.res-legal.eu/compare-support-schemes/>.

69 Jonathan A Lesser and Xuejuan Su, 'Design of an Economically Efficient Feed-in Tariff Structure for Renewable Energy Development' (2008) 36 *Energy Policy* 981.

70 Couture et al., above n 37, 955; Davies, 'Incentivizing Renewable Energy Deployment', above n 40, 56; Peng Sun and Pu-yan Nie, 'A Comparative Study of Feed-in Tariff and Renewable Portfolio Standard Policy in the Renewable Energy Industry' (2015) 74 *Renewable Energy* 255, 261.

71 Hans-Josef Fell, '*Feed-in Tariff for Renewable Energies: An Effective Stimulus Package Without New Public Borrowing*' (2009) Deutscher Bundestag 20.

72 Sovacool, *Renewable Electricity for Southeast Asia*, above n 3, 23–5.

73 Davies, 'Reconciling Renewable Portfolio Standards and Feed-in Tariffs', above n 2, 343–4; Parliamentary Library of the Commonwealth of Australia, above n 39.

74 Sawin, above n 3, 5.

75 Ingmar Ritzenhofen and Stefan Spinler, 'Optimal Design of Feed-in-Tariffs to Stimulate Renewable Energy Investments Under Regulatory Uncertainty-A Real Options Analysis' (Research Paper, Social Science Research Network, 29 April 2013) 5; Maria Ellingson et al., *Compendium of Best Practices: Sharing Local and State Successes in Energy Efficiency and Renewable Energy from the United States* (REEEP/ACORE, 2010) 59–60.

sources and technologies.[76] It is also difficult to predict in advance how much energy will actually be generated under a FIT.[77] In a number of countries these two factors have led to significant cost blowouts, which have led to FITs being cancelled early or even amended retrospectively.[78] Further, international experience has repeatedly shown that it can be time consuming and difficult to adjust tariff levels quickly. When a tariff is amended retrospectively or cancelled, confidence in the stability of the market diminishes and it becomes a more risky investment proposition. It is for these reasons that the EU State Aid Guidelines have stated that feed-in tariffs are no longer to be supported as an eligible form of state aid by Member States, with a preference for competitive tendering and feed-in premiums instead.[79]

5.2.2 *Feed-in Premiums*

Feed-in premiums (FIPs) operate on a similar basis to feed-in tariffs, but are a market-dependent regulatory support mechanism rather than a market-independent one. FIPs provide either a fixed or variable premium (also sometimes called an 'environmental bonus')[80] above the spot price for electricity for each kWh of renewable electricity generated and fed into the grid. Theoretically, the premium is set at a level equal to the externalities associated with conventional fossil fuel sources of generation. However, Haas et al. have reported that due to the difficulties associated with pricing those externalities, in reality, many fixed premiums are based on the estimated production costs of renewable energy when compared with the electricity price rather than any environmental benefits.[81] As with FITs, FIPs are usually guaranteed for a period of ten to twenty years.[82] However, Lithuania and Hungary uses predetermined production caps for each technology to prevent cost blowouts.[83]

76 Parliamentary Library of the Commonwealth of Australia, above n 39.

77 Wüstenhagen et al., above n 53, 7–8.

78 Thomas Gerke, *Italy Imposes Retroactive Changes to Feed-in Tariff for Solar PV* (15 August 2014) RE New Economy <http://reneweconomy.com.au/2014/italy-imposes-retroactive-changes-feed-tariff-pv-38857>; Nilima Choudhury, *Spain Announces Retroactive FiT Cuts* (19 February 2013) PV Tech <www.pv-tech.org/news/spain_announces_retroactive_fit_cuts>; James Martin, *Western Australia Announces Retroactive Feed-in Tariff Cuts* (9 August 2013) Solar Choice <www.solarchoice.net.au/blog/news/western-australia-announces-retroactive-feed-in-tariff-cuts-090813/>.

79 European Commission, *Guidelines on State Aid for Environmental Protection and Energy 2014–2020*, (2014) C200/1, 28.6.2014, Section 3.3 Aid to energy from renewable sources.

80 Frässe-Ehrfeld, above n 13, 264.

81 Haas et al., 'A Historical Review of Promotion Strategies', above n 1, 1011.

82 Frässe-Ehrfeld, above n 13, 264.

83 RES Legal, *Compare Support Schemes* (2018) <www.res-legal.eu/compare-support-schemes/>.

5.2.2.1 Fixed Premium Price Model

There are three different models of FIPs: the fixed premium price model, the variable premium price model and the percentage of retail price model.[84] Under the fixed premium price model, renewable electricity is sold into the spot market rather than under long-term power purchase agreements. As a result, renewable energy generators are remunerated by a combination of the variable market price and a fixed premium price. The fixed premium may be banded according to technology type and/or project size. The fixed premium price model has been criticised as imposing greater risks than other models 'that payment levels will either be too high, or too low, which can have negative consequences for market growth, investor security, and for society at large'.[85] In addition, many premium price schemes do not have a purchase obligation, putting added pressure on renewable electricity generators. These factors have led to FIPs being, on average, more costly per kWh than FITs.[86] The additional cost for FIPs has been attributed to project developers needing additional compensation for the added risk and 'the greater likelihood of divergence between total remuneration and actual project costs'.[87] The fixed premium price model does, however, offer some benefits; for example, FIPs allow for renewable generation to be better integrated into competitive electricity markets. In addition, they provide an incentive for renewable electricity generators to generate during peak periods of demand in order to achieve higher prices.[88]

5.2.2.2 Variable Premium Price Model

The variable premium price model (also called the 'sliding price' model)[89] introduces both caps and floors to the prices in order to provide greater investor security when electricity prices decline markedly, and to prevent windfall profits in the event that electricity prices rise suddenly.[90] As electricity prices rise, the premium tails off until it reaches the cap, at which point the premium price is zero and the renewable electricity generator receives the electricity spot price. Couture and Gagnon argued that this

84 Couture et al., above n 37, 960–2.
85 Ibid 960.
86 Julieta Schallenberg-Rodriguez and Reinhard Haas, 'Fixed Feed-in Tariff Versus Premium: A Review of the Current Spanish System' (2012) 16 *Renewable and Sustainable Energy Reviews* 293, 304.
87 Couture et al., above n 37, 960.
88 Ibid.
89 IRENA, above n 47, 10.
90 Kitzing et al., above n 44, 195.

model is significantly riskier than other regulatory support mechanisms due to the fact that the remuneration of the renewable electricity generator is subject to market volatility, including that resulting from the use of conventional fossil fuel sources, which is beyond the control of the renewable electricity generator.[91]

Contracts for difference, which are used in the United Kingdom and some of the Australian states, are variants on the fixed premium price model. Unlike standard FIPs, which relate to the actual supply of electricity, CFDs are derivatives, whereby the renewable generator is paid the difference between the contractually agreed guaranteed rate of revenue or 'strike-price', and the market price for electricity. The strike-price is often set through a reverse auction. CFDs can be unilateral hedges. However, they are more commonly bilateral hedges, with the generator receiving payment when the market price is less than the strike price but also having to pay either all or some of the difference between the market price and the strike price when the market price is higher.[92]

5.2.2.3 Percentage of the Retail Price Premium Price Model

The percentage of the retail price model is the last premium price model to be considered. This model provides renewable electricity generators with a fixed percentage of the retail electricity price for the sale of their renewable electricity. The percentage can establish that the purchase price is either above, equal to or beneath the average retail price for electricity in a given market. This model is no longer used, with countries such as Germany switching to a fixed-price FIT as a way of increasing investor security and providing greater stability to the renewable energy sector.[93]

5.2.3 *Renewable Portfolio Standards with Tradeable Green Certificates*

Renewable portfolio standards (also known as a 'quota obligation') operate on the basis of a quota system, creating a legal obligation on licensed electricity suppliers to supply a specified percentage or volume of their

[91] Couture et al., above n 37, 961.

[92] Melbourne Renewable Energy Project, 'Which Model Is the Right Model? A Guide to Renewable Energy Procurement' 6 <http://melbourne.vic.gov.au/mrep>; See generally Phillip Wild, 'Determining Commercially Viable Two-way and One-way "Contract-for-Difference" Strike Prices and Revenue Receipts' (2017) 110 *Energy Policy* 191.

[93] Ibid.

electricity from renewable energy generators.[94] The quota is often set nationally to encourage least cost deployment of renewables, however, India uses different quotas for each state, and China has proposed the adoption of provincial quotas in their draft RPS.[95] The quota generally increases over time, with specific targets to be achieved at regular intervals, and a final target to be achieved by the expiry date of the renewable portfolio standard. Renewable portfolio standards are currently used in thirty countries around the world.[96]

As with many of the regulatory support mechanisms, RPS programmes show significant variability in their design, compliance mechanisms, technologies supported and administration. A commonality is that most countries require that their licensed electricity suppliers prove that they have met their quota obligation by presenting one tradeable green certificate[97] ('TGC') for each MWh[98] of their quota of renewable electricity.

TGCs are electronic certificates issued to certify that electricity was generated from renewable energy sources. They are a separate commodity, distinct from the renewable electricity generated. In many countries, TGCs are able to be traded separately from the actual electricity produced, although some countries insist that both the renewable electricity and the TGC are bundled together.[99] The advantage of allowing the TGCs to be traded separately from the physical electricity generated is that it provides more flexibility to electricity suppliers by removing:

94 See generally Leon et al., 'Designing the Right RPS', above n 3; Berry et al., above n 3; Ellingson, above n 76; Davies, 'Incentivising Renewable Energy Deployment', above n 40; Sawin, above n 3.

95 国家能源局综合司关于征求《可再生能源电力配额及考核办法（征求意见稿）意见的函 [Draft Letter from the General Department of the National Energy Administration on the Opinions on Soliciting the Renewable Energy Power Quota and Assessment Measures, National Energy Board, 23 March 2018].

96 REN21 Secretariat, 'Renewables 2018 Global Status Report', above n 38, 64–7.

97 TGCs are also sometimes referred to as renewable energy certificates (RECs), renewable energy credits, green tags, green credits, green electricity certificates or certificates of origin: The Energy and Resources Institute, *Renewable Energy Credits: Prevailing Practices* (Report No. 2005RT24, REEEP Project, 2006) 1–2.

98 Though it has been reported that in some TGC schemes one TGC is issued for each kWh: see e.g. Thomas P Lyon and Haitao Yin, 'Why Do States Adopt Renewable Portfolio Standards? An Empirical Investigation' (2010) 31 *The Energy Journal* 131, 133.

99 Leon et al., 'Designing the Right RPS', above n 3, 10, 30–1; Berry et al., above n 3, 267; The Energy and Resources Institute, *Renewable Energy Credits*, above n 95, 10; Grace et al., above n 3, 16.

> geographical or physical limitations (such as resource availability), time-scale (such as seasonal availability or mismatch between supply and demand) or financial limitations associated with the supply of renewable sources.[100]

TGCs are often used as a means of tracking and verifying that the correct quantities of renewable electricity have been generated to meet an electricity supplier's obligation under the quota.[101]

Under a traditional RPS, the quota is technology neutral, with each MWh of electricity generated from an eligible renewable energy source awarded one TGC. The theory behind not imposing conditions on the types of technologies to be supported is that it would mean that regulators were not involved in 'picking winners' and would therefore ensure that the quota obligation was met at least cost. This usually means that the quota is met by renewable electricity produced by large renewable generators that have economies of scale, and use the most commercialised technologies.[102] Many countries that were early adopters of the RPS, such as the United Kingdom, Italy and Belgium, found that the technology neutral structure of the traditional RPS failed to provide them with a sufficient diversity of supply to meet their energy security needs.[103] As a result, a variant (the 'modern RPS') was devised which uses carve-outs and/or technology banding with credit multipliers in order to give preferences to some technologies or project types over others. Carve-outs or set-asides are a means of setting different targets within an RPS for different technologies or project types.[104] An example of this is the large-scale renewable energy target (LRET) and the small-scale renewable energy target (SRET) in Australia, each of which has its own eligibility criteria and rules.[105]

A more common approach seems to be the use of technology banding with credit multipliers. As stated above, under a traditional RPS, 1MWh of renewable electricity is assigned one TGC. Under a credit multiplier approach this is varied with 1MWh of renewable electricity produced by specific renewable energy sources or technologies being credited a positive or negative multiple of one TGC, depending on its relative degree of

100 Mitchell, *The Political Economy of Sustainable Energy*, above n 4, 19.

101 Leon et al., 'Designing the Right RPS', above n 3, 10; The Energy and Resources Institute, *Renewable Energy Credits*, above n 98, 9.

102 Mitchell, *The Political Economy of Sustainable Energy*, above n 4, 12–13.

103 Verbruggen et al., above n 7, 638–40.

104 Leon et al., 'Designing the Right RPS', above n 3, 40–1.

105 *Renewable Energy (Electricity) Act 2000* (Cth).

commercialisation.[106] For example, in the 2018 obligation period in South Korea, 1MWh of electricity generated from landfill gas was credited 0.5 Renewable Energy Certificates (RECs) (their equivalent of a TGC), whereas hydrogen fuel cells and tidal power which were less commercialised were credited two RECs per MWh.[107]

There are commonly three different strategies (which may either be used in isolation or in combination) that licensed electricity suppliers can use to meet their quota of TGCs under the RPS:

1. they can generate the specified percentage or volume of electricity from renewable energy sources themselves (i.e. they themselves produce and keep sufficient renewable electricity and TGCs to meet their quota); and/or
2. they can purchase electricity generated by a third party from renewable energy sources. This electricity can then be used to meet the company's obligation under the RPS, with the electricity then on-sold to their final customers (i.e. they purchase sufficient renewable electricity and the accompanying TGCs from a third party to meet their quota); and/or
3. they can purchase TGCs to meet their obligation.

In most countries, if the licensed electricity supplier fails to meet their quota obligation in the relevant obligation period, they must pay a specified amount per TGC, 'the buyout price', into a buy-out fund. In South Korea, the buy-out price is calculated as 150 per cent of the average annual REC transaction price for their REC shortfall.[108] The buy-out price is usually re-assessed annually and in some countries, such as the United Kingdom, it is indexed to the Retail Prices Index.[109] Payment of the buy-out price effectively acts as a fixed penalty for each TGC shortfall and allows the electricity suppliers to discharge in whole or in part their obligations under the RPS.[110]

At the end of each obligation period, the government entity responsible for managing the buy-out fund normally distributes it (along with interest earned on the principal) among all of the electricity suppliers who have complied, either in whole or in part, with the RPS by presenting TGCs to the responsible

106 Leon et al., 'Designing the Right RPS', above n 3, 41–2.

107 Korean Energy Agency, *Renewable Portfolio Standards* (2018) <www.energy.or.kr/renew_eng/new/standards.aspx>.

108 Ibid.

109 Penelope Crossley, Miles Curley and John Pickett, 'Legislación sobre energías renovables en el Reino Unido' in Fernando Becker et al. (eds.), *Tratado De Energias Renovables: Volumen II. Aspectos jurídicos* (Thomson Aranzadi/Iberdrola, 2010).

110 Ibid.

entity.[111] Each supplier is awarded a share of the funds proportionate to the ratio of TGCs it has produced compared to the total amount of TGCs received by the relevant entity over the obligation period. In this way, revenue from non-compliant suppliers is fed back to compliant suppliers.[112] In practice, it is usual for a large portion of the buy-out payment to be passed back to the renewable generator under the terms of the renewable power purchase agreement (PPA).[113] As TGCs are freely tradeable between renewable generators and licensed electricity suppliers their value is driven by the forces of supply and demand. The greater the shortfall between the quota obligation and actual renewable generation, the higher the TGC buy-out price becomes. As the TGC buy-out price increases, so does the incentive to invest in new renewable generation.[114]

Where the licensed electricity supplier has repeatedly failed to meet their quota of TGCs and has also failed to pay the buy-out price, this is likely to breach the conditions of their operating licence. Ultimately, if this situation is not rectified, the licensed electricity supplier may have their operating licence suspended, revoked or not renewed for future terms.[115]

Some countries do allow a degree of flexibility towards electricity suppliers to assist them in meeting their quota obligations under the RPS. There are three flexibility mechanisms that can be incorporated into an RPS: TGC banking, TGC borrowing and compliance waivers. TGC banking involves the purchase of excess TGCs in years when there is a surplus, which are then banked to use to meet their obligation in a future year.[116] TGC borrowing permits electricity suppliers who have a TGC shortfall in a given obligation period to defer the shortfall to the following year.[117] Compliance waivers are where an electricity supplier that is going to have a shortfall of TGCs, due to being unable to purchase sufficient renewable energy to meet their obligation, requests permission from the managing government entity to not have to comply with their obligation in that obligation period.[118] Compliance waivers are usually issued on a one-off basis in years where there is 'a significant shortage of renewable energy generation beyond the utilities' control'.[119] While TGC banking is arguably to the advantage of renewable energy

111 Ibid.
112 Ibid.
113 Ibid.
114 Ibid.
115 Ibid.
116 Leon et al., 'Designing the Right RPS', above n 3, 32–3.
117 Ibid 33–4.
118 Ibid 34–5.
119 Ibid 34.

generators, neither TGC borrowing nor compliance waivers operate to their advantage. The latter two flexibility mechanisms make the operation of the RPS less stable and provide less certainty as to the quantity of electricity generated from renewable energy sources in any given obligation period. This may lead to renewable energy generators seeking to defer their projects, electricity suppliers focusing their attention on seeking waivers rather than complying with their obligation and increase the administrative burden on the responsible government entity.[120]

5.2.3.1 The Advantages and Disadvantages of Using Renewable Portfolio Standards with Tradeable Green Certificates

The key advantage of a traditional RPS is that the government sets a quota for the desired level of renewable generation within the country, but it is then left to market forces to decide the mix of renewable energy sources and technologies used to meet that quota, as well as the price paid. There is flexibility in this approach, as the licensed electricity supplier can use one of the three approaches detailed above to meet their obligation. If the licensed electricity suppliers are behaving rationally under an RPS they will select the cheapest form of renewable generation, which in turn will reduce the costs borne by the end-consumers. However, this approach has been criticised for ignoring the 'qualification of RE supplies, promoting already mature and less sustainable RE supplies while neglecting more promising sources that are not quite as close to market-readiness'.[121]

In the case of a modern RPS, the use of carve-outs and/or technology banding with credit multipliers means that RPS are no longer technologically neutral and that governments are now engaged in 'picking winners'. While a diverse portfolio of renewable energy sources and technologies is beneficial in terms of ensuring energy security, its efficiency and cost-effectiveness is now dependent on the government being able to accurately select the correct mix for the market. Further, the complexity involved in the design of the modern RPS may also make the implementation, and monitoring and compliance, more technically difficult and costly. Indeed, repeated studies have shown that RPS have failed to deliver the same quantities of deployment as FITs and that the average cost of that deployment has also been higher.[122] This reflects that in the case of both the traditional and modern RPS, the lack of ongoing

[120] Ibid 32–5.

[121] Verbruggen et al., above n 7, 642.

[122] Julieta Schallenberg-Rodriguez, 'Renewable Electricity Support Systems: Are Feed-in Systems Taking the Lead?' (2017) 76 *Renewable and Sustainable Energy Reviews* 1422; Peng Sun and Pu-yan Nie, 'A Comparative Study of Feed-in Tariff and Renewable

certainty over the price to be paid increases the risk and uncertainty for investors.[123] Concerns have also been expressed that the quota may act as an 'unintentional ceiling on renewable energy development, with little incentive to go beyond the minimum rate set by the policy'.[124]

Many of these problems can be overcome through properly designed RPS that provide long-term stability, yet for which compliance can be simply monitored. However, the challenges associated with RPS mean that they may not be suitable for all countries, which likely explains why the popularity of RPS has diminished in recent years in favour of FITs, FIPs and competitive tendering processes.[125]

5.2.4 *Competitive Tendering and Auction Bidding*

Under competitive tendering processes, the government issues a request for proposal (RFP) for the installation of a specific capacity of electricity generated from renewable sources (kW), a specific volume of generation (kWh) or announces a set budget.[126] Depending on the desired outcomes of the country, the RFP can be technology-neutral (least cost of deployment but may also lead to windfall profits for highly commercialised sources) or technology-specific.[127] During the RFP process, applicants are required to submit a comprehensive 'sealed-bid' proposal that details the technical, economic, environmental and financial details of the proposed project.[128] Applicants may also be required to show that they have met certain pre-qualification requirements such as having

Portfolio Standard Policy in Renewable Energy Industry' (2015) 74 *Renewable Energy* 255; C G Dong, 'Feed-in Tariff vs. Renewable Portfolio Standard: An Empirical Test of Their Relative Effectiveness in Promoting Wind Capacity Development' (2012) 42 *Energy Policy* 476; Haas et al., 'A Historical Review of Promotion Strategies', above n 1, 1026; Lucy Butler and Karsen Neuhoff, 'Comparison of Feed-in Tariff, Quota and Auction Mechanisms to Support Wind Power Development' (2008) 33 *Renewable Energy* 1854, 1858; Couture et al., above n 37, 955; Lakshmi Alagappan, Ren Orans and Chi-Keung Woo, 'What Drives Renewable Energy Development?' (2011) 39 *Energy Policy* 5099, 5099.

123 Katrin Jordan-Korte, *Government Promotion of Renewable Energy Technologies: Policy Approaches and Market Development in Germany, the United States, and Japan* (Gabler Research, 2011) 138.

124 Benjamin K Sovacool, 'A Comparative Analysis of Renewable Electricity Support Mechanisms for Southeast Asia' (2010) 35 *Energy* 1779, 1786.

125 Mir-Artigues et al., above n 5, 434.

126 Pablo del Río, 'Designing Auctions for Renewable Electricity Support. Best Practices from Around the World' (2017) 41 *Energy for Sustainable Development* 2.

127 See e.g. Erik Gawel et al., 'Rationales for Technology-specific RES Support and Their Relevance for German Policy' (2017) 102 *Energy Policy* 16.

128 CESA, *Developing an Effective State Clean Energy Program: Competitive Grants* (CESA, 2009) 1.

obtained planning permission, preliminary licences or other technical requirements. The RFPs are then reviewed by the programme managers in accordance with a competitive framework to determine whether the project will be supported and the level of funding to be provided.[129]

An alternative method of allocating competitive grants is a reverse auction bidding process. Using this method, the government 'defines a reserve market for a given amount of [electricity generated from renewable energy sources] and organises a competition between renewable producers to allocate this amount'.[130] The reverse auction is usually conducted using a standard form contract, with a few terms that may be the subject of bidding such as the price, quantity, delivery dates and minimum performance standards. As with all government procurement, while the criteria against which the tenders may be judged will vary, in most cases the bids that best meet the government's requirements with the least cost proposed per kWh during the bidding process will be selected.[131] In order to meet the required amount of renewable electricity under the reverse auction, 'the proposals are classified in increasing order of cost until the amount to be contracted is reached'.[132] At the conclusion of the auction, each of the successful bidders is awarded a long-term contract on the terms of the standard contract to supply electricity generated from renewable sources at their bid price.

One of the first reverse auction processes for renewable energy was introduced in the United Kingdom under the Non-Fossil Fuel Obligation (NFFO). The NFFO was established on 1 October 1990 to provide a subsidy to the State-owned nuclear companies, following the privatisation of the rest of the electricity generation sector in the United Kingdom in 1989.[133] The NFFO later came to be used by the renewable energy sector, with a plan to deliver 1500MW of installed capacity from renewable energy sources by 2000.[134] The NFFO tender process sought to support generation from a range of renewable energy technologies, with certain quantities of installed capacity to be realised by different renewable energy technologies in the form of a reverse auction.[135] Those project developers awarded contracts were given long-term contracts of up to fifteen years' duration, with:

129 Ibid.

130 Menanteau et al., above n 1, 802.

131 Vasilios Anatolitis and Marijke Welisch, 'Putting Renewable Energy Auctions into Action – An Agent-based Model of Onshore Wind Power Auctions in Germany' (2017) 110 *Energy Policy* 394, 395.

132 Menanteau et al., above n 1, 802–3.

133 Mitchell, *The Political Economy of Sustainable Energy*, above n 4.

134 Haas et al., 'Promoting Electricity from Renewable Energy Sources', above n 3, 441.

135 Sawin, above n 3, 6–7.

> . . . a guaranteed surcharge per unit of output for the entire contract period. The difference between the surcharge paid to NFFO generators (premium price) and a reference price (Pool Selling Price) was to be financed by a levy on all electricity sales of licensed electricity suppliers. The costs of this levy were to be passed on to consumers.[136]

Five bidding rounds were conducted in England and Wales under the NFFO, with 880 contracts being awarded.[137] Over this period, the average price per kWh of renewable energy dropped considerably. In particular, in Scotland, the price paid for wind power per kWh dropped to a point at which it was lower than that for electricity generated from coal, oil, nuclear and even some natural gas sources.[138] These price reductions may be related to a number of factors such as declining technology costs, improved site selection, learning experience of operators, better economies of scale and improved technical performance.[139]

Despite this apparent success, the general consensus among scholars is that the NFFO was profoundly flawed.[140] Many of the projects awarded contracts were never constructed, while others failed to meet their contractually agreed levels of installed capacity. Indeed, less than a third of the contracts awarded for wind power projects were ever realised.[141] Thus, across the range of renewable energy technologies there were shortfalls in the levels of installed renewable energy generation capacity to be achieved by the NFFO. This failure has been primarily attributed to bidders, under the pressure of stiff competition, submitting aggressive bids containing unrealistic bid prices in an attempt to secure a contract.[142] The lack of effective penalties for non-delivery meant that there were no real costs involved in not fulfilling their contractual obligation.[143] A further problem was that at the time of bidding, project developers were not required to seek prior planning consent, leading to

[136] Haas et al., 'Promoting Electricity from Renewable Energy Sources', above n 3, 441–2.

[137] Ibid 442.

[138] Menanteau et al., above n 1, 807–8; Haas et al., 'Promoting Electricity from Renewable Energy Sources', above n 3, 442.

[139] Menanteau et al., above n 1, 808.

[140] Catherine Mitchell and Peter Connor, 'Renewable Energy Policy in the UK 1990–2003' (2004) 32 *Energy Policy* 1935, 1936–8; Haas et al., 'Promoting Electricity from Renewable Energy Sources', above n 3, 442, 444; Menanteau et al., above n 1; Lesser et al., above n 70, 983.

[141] Lesser et al., above n 70, 983; see also Sovacool, *Renewable Electricity for Southeast Asia*, above n 3, 22.

[142] Ottinger et al., above n 2, 199; Alagappan et al., above n 124, 5101; Sovacool, *Renewable Electricity for Southeast Asia*, above n 3, 22.

[143] Mitchell et al., above n 141, 1937; Haas et al., 'Promoting Electricity from Renewable Energy Sources', above n 3, 442; Lesser et al., above n 70, 983; Menanteau et al., above n 1, 806–7.

some projects being subsequently denied planning permission, while other project developers had problems trying to interconnect to the transmission lines.[144] As a result, the NFFO programme was closed and replaced in April 2002 by a renewable portfolio standard or quota obligation scheme, the Renewables Obligation Order.[145] This pattern seems to have been reflected in the experience of other countries, with similar schemes adopted during the 1990s in France, Ireland, Denmark, Scotland, Northern Ireland and many states in the United States.[146] Most of these schemes were abandoned in the early 2000s.[147]

Similar issues were also experienced in the early rounds of the Chinese wind power concession auctions in 2003–4. These auctions were characterised by inexperienced bidders bidding at levels below marginal investment costs and thus the non-fulfilment of some of the contracts issued. Rather than scrapping the scheme, the Chinese Government learnt from its previous successes and failures. It adjusted the auction design for each subsequent round, including by introducing, and subsequently adjusting the weighting on, strict prequalification criteria and imposing penalties for non-performance of contracts issued to winning bidders.[148]

Many countries have now adopted this approach, with auctions viewed as an attractive option for both high-income countries and low-income countries due to their flexibility and promise of least cost deployment. Modern competitive tendering schemes are now designed to overcome many of the earlier problems. In particular, appropriate penalties are often now included to prevent non-delivery of projects.[149] Further, most countries have now changed their national planning policy and laws to give renewable energy project developers greater certainty that their planning application will be approved.[150]

This has prompted a resurgence in competitive tendering schemes. They are now the most common regulatory support mechanism, with eighty countries, including Brazil, France, Mexico, Indonesia, Russia, Thailand, South Africa and the United States having adopted competitive tendering or auction

144 Haas et al., 'Promoting Electricity from Renewable Energy Sources', above n 3, 442, 444.
145 Mitchell et al., above n 141.
146 Haas et al., 'Promoting Electricity from Renewable Energy Sources', above n 3, 441. See also Menanteau et al., above n 1, 802.
147 Ibid.
148 Xiaodong Wang et al., 'Promoting Renewable Energy Through Auctions: The Case of China' *Livewire World Bank Group* (2014/14), 888697.
149 European Commission, *Guidance for the Design of Renewables Support Schemes*, SWD (2013) 439 Final (5 November 2013) 6.
150 See e.g. Department of Energy and Climate Change, 'National Policy Statement for Renewable Energy Infrastructure' (Paper, United Kingdom Parliament, July 2011).

processes by early 2018.[151] Competitive tendering is also now the preferred regulatory support mechanism for EU Member States, with the current State Aid Guidelines specifying that:

> Market instruments, such as auctioning or competitive bidding process open to all generators producing electricity from renewable energy sources competing on equal footing at EEA level, should normally ensure that subsidies are reduced to a minimum in view of their complete phasing out. However, given the different stage of technological development of renewable energy technologies, these Guidelines allow technology specific tenders to be carried out by Member States, on the basis of the longer-term potential of a given new and innovative technology, the need to achieve diversification; networks constraints and grid stability and system (integration) costs.[152]

Competitive tendering is seen as an effective regulatory support mechanism in terms of both its cost effectiveness[153] and its fiscal responsibility, as the costs of running such a scheme are passed on to electricity consumers.[154] Further, del Río has stated that 'auctions mitigate the information asymmetry problem when setting remuneration levels, [and auctions] are particularly suitable to control costs, expansion and technology mix and they are more likely to lead to allocative efficiency'.[155] The efficiency argument has also been made by Leon, who has argued that reverse auctions can be an efficient way of delivering renewable energy because they enable 'the level of incentive to be set by the lowest-cost renewable projects, while not paying more than necessary'.[156] This has been the experience in the wind sector in Brazil, where the use of reverse auctions coupled with long-term power purchase agreements led to a cost reduction of 42 per cent when compared to the previously used FIT rates.[157] However, experience suggests that if no specific provision is made for different technologies, competitive tendering will not support the development of less-established technologies that may potentially provide more

151 REN21 Secretariat, 'Renewables 2018 Global Status Report', above n 38, 64–7.

152 European Commission, *Guidelines on State aid for environmental protection and energy 2014–2020*, (2014) C200/1, 28.6.2014, Section 3.3 Aid to energy from renewable sources, paragraphs 109–10.

153 Ottinger et al., above n 2, 199; Sovacool, *Renewable Electricity for Southeast Asia*, above n 3, 22; The Energy and Resources Institute, above n 22, 11

154 Sovacool, *Renewable Electricity for Southeast Asia*, above n 3, 22–3.

155 Pablo del Río, 'Designing Auctions for Renewable Electricity Support. Best Practices from Around the World' (2017) 41 *Energy for Sustainable Development* 1.

156 Leon et al., 'Designing the Right RPS', above n 3, 52.

157 OECD, *OECD Policy Guidance for Investment in Clean Energy Infrastructure* (OECD, 2013) 20.

efficient and cost-effective delivery of renewable energy in the future.[158] This point is identified by Sovacool, who has argued that competitive tendering lacks static efficiency, does not encourage dynamic efficiency and is also inequitable.[159] He argues that competitive tendering schemes usually favour large incumbent players in the market, including state-owned enterprises (SOEs) (where they still exist) without profit motives.[160] For example, independent power providers and small firms may choose not to participate if they believe that their chance of submitting a winning bid is much lower because they cannot show a strong track record or the economies of scale.[161] Indeed, where these schemes have existed without penalties in place for non-delivery, there has been evidence that 'a small number of players have "gamed" bid prices to block out competitors (but never intended to achieve complete projects)'.[162] A further problem occurs where private firms are forced to compete with SOEs that may not have the same profit motives, and can therefore commit to unreasonably low prices.[163] del Río and Linares have also identified that depending on the structure, size and lead times of the auctions, they can also be beset by high transaction and administrative costs if these factors are not appropriately managed.[164]

It is perhaps too early to be able to evaluate the effectiveness of the latest round of competitive tendering programmes. However, due to their focus on the least cost delivery of renewable energy, if some of the other problems can be overcome, competitive tendering programmes may be an effective method of rapidly deploying the technologies that are currently the least-cost option.

5.2.5 *Net Metering*

Net metering is another popular regulatory support mechanism that has been adopted by fifty-five countries by 2018, including Denmark, Latvia, Sri Lanka and Tanzania. It is most commonly used to encourage the deployment of small-scale renewable energy projects amongst residential and small business customers. Net metering involves customers who have a source of renewable energy generation installed, commonly photovoltaic solar cells, connecting it

158 Ibid.
159 Sovacool, *Renewable Electricity for Southeast Asia*, above n 3, 22.
160 Ibid.
161 Leon et al., 'Designing the Right RPS', above n 3, 52.
162 Sovacool, *Renewable Electricity for Southeast Asia*, above n 3, 22.
163 Ibid.
164 Pablo del Río and Pedro Linares, 'Back to the Future? Rethinking Auctions for Renewable Electricity Support' (2014) 35 *Renewable and Sustainable Energy Reviews* 42, 50.

to the electricity distribution network using a bi-directional electricity meter. The bi-directional meter registers when the customers produce electricity that is surplus to their own needs and it gets exported to the grid, and also, conversely, it registers when they have an electricity deficit and so have to import electricity from the grid.

The payment for the exported electricity is then credited against the cost of the imported electricity supplied by the utility company. At the end of each billing period, the customer is charged the net cost of the electricity (that is, the cost of electricity imports less the cost of electricity exports). Where the net cost of the electricity used by the customer is negative, the customer is credited this amount as against their next electricity bill.[165] The customer normally also retains the ownership of any environmental benefits attained through their generation of renewable electricity such as renewable energy credits/TGCs.

Sawin has stated that net metering offers advantages to both network operators and electricity generators by improving system load factors and offsetting the need for new peak load generating plants. However, drawing upon the experiences of Texas and California, she also notes that without other financial incentives, net metering alone will not advance market penetration of renewable energy.[166] Verbruggen and Lauber agree with this analysis, finding that because less competitive technologies such as small onshore wind turbines or PV solar currently have generation costs that exceed the retail electricity prices, providing remuneration at retail price levels will not be sufficient to stimulate growth in these technologies.[167] They argue that if the purpose of net metering is 'to stimulate technological development and learning, remuneration should be based on generated renewable energy quantities (irrespective of whether used on site or delivered to the grid)'.[168]

A further challenge to net metering is the disruptive nature of the new energy storage technologies currently being developed and commercialised.[169] As energy storage technologies become more cost-competitive and commercially available, net metering is likely to decline as the amount paid for electricity exported to the grid is often significantly less than the cost of electricity imported. Further, with the increasing penetration of energy storage technologies, the advantages of price arbitrage will diminish, leading to a long-term diminution of the price spread between the peak and off-peak

165 Ernest E Smith, 'US Legislative Incentives for Wind-Generated Electricity: State and Local Statutes' (2005) 23 *Journal of Energy & Natural Resources Law* 173, 180.
166 Sawin, above n 3, 5.
167 Verbruggen et al., above n 7, 637.
168 Ibid.
169 UBS, 'We Love a Sunburnt Country' (Report, UBS Utilities Sector, 17 May 2014) 2.

periods.[170] In these circumstances, it may become more cost-effective to store excess electricity generated and then use it when renewable energy cannot be generated, such as at night time (for photovoltaic solar) or when wind is not blowing (for wind energy).

5.2.6 *Renewable Energy Targets*

Renewable energy targets (RETs) are commitments or goals set by government that stipulate that either a specific percentage or volume of installed capacity or generation will be met from renewable energy by a future date. The targets set for the percentage of electricity to be generated from renewable sources vary significantly. For example, RETs for 2030 range from 5 per cent for Bahrain (up from 0.2 per cent of installed generating capacity in 2015) to 100 per cent in Costa Rica, Fiji, Papua New Guinea, Samoa and the Solomon Islands.[171] The targets may be legislated or established under policies set by the relevant ministries or energy authorities.

RETs are currently the most popular form of policy intervention within the renewable energy sector, existing in 146 countries in early 2018.[172] There are many variations of the key features of RETs, including whether:

1. the targets are legally binding or non-binding aspirational goals;
2. the targets focus on renewable electricity generation alone or are focused on the heat and transport sectors either individually or through a combined target;
3. the target is based on the renewable share of primary energy, final energy, the installed capacity of particular renewable energy sources/technologies or their energy output; and
4. there is a set date for the achievement of the target or no end date.

Where the RET is legally binding it will normally be coupled with a TGC scheme to make monitoring and compliance easier. However, many countries have non-binding targets. The success or failure of a RET depends on whether it has been appropriately designed in the context of the domestic market to which it applies.

170 See e.g. Bartholomäus Wasowicz et al., 'Evaluating Regulatory and Market Frameworks for Energy Storage Deployment in Electricity Grids with High Renewable Energy Penetration' (Paper presented at the 9th International Conference on the European Energy Market, Florence, 2012) 2.

171 REN21 Secretariat, 'Renewables 2018 Global Status Report', above n 38, 189–91.

172 Ibid 53.

A number of countries that were relatively early movers with RETs have reviewed their RETs in recent years to decide whether they should continue to operate in their current form. In the United Kingdom, the government has announced that there will not be a RET beyond 2020, with their domestic mix of low-carbon energy to be made up of a mix of sources including nuclear power, natural gas and renewable energy to be decided on a technology-neutral basis by the market.[173] The EU-wide target of at least 32 per cent of energy to be generated from renewable sources by 2030 will continue to apply. However, unlike the 2020 target that was enforceable for each Member State, the 2030 target will not be binding on individual Member States, with the target being for the EU as a whole.[174] There has already been speculation that this means that in the EU after 2020 'there will be no meaningful renewable energy target'.[175]. In 2014, Australia reviewed its national RET with the Review Panel chaired by Dick Warburton, the former Chairman of Caltex Oil in Australia. This Review recommended that the LRET be changed from a volumetric target of 41,000GWh of electricity from large-scale renewable energy by 2020 to a market share target of 20 per cent of electricity generation by 2020.[176] This reflected industry concerns that, in the context of declining electricity demand and greater energy efficiency, the volumetric requirement of the RET meant that approximately 27 per cent of electricity generation would have been derived from renewable energy sources if the RET were to be met.[177] This was significantly higher than the 20 per cent originally intended when the RET was designed. The Review recommended that, to protect existing generators, the RET be revised to a 'real 20 per cent target' for large-scale renewable generation (equivalent to approximately 33,000GWh), rather than using the current 41,000GWh production target. The Review

173 Department of Energy and Climate Change, United Kingdom, 'A 2030 Framework for Climate and Energy Policies: UK Government Response to Commission Green Paper' COM(2013) 169 final, 1 July 2013, 8–9. Fiona Harvey, 'Loss of renewable target is backward step in fight against climate change', *The Guardian* (online), 23 January 2014 <www.theguardian.com/environment/2014/jan/22/no-renewable-target-climate-change>; Tom Bawden, 'EU Admits It Has No Power to Enforce Its 'Binding' 2030 Renewable Energy Targets', *The Independent* (online), 22 January 2014 <www.independent.co.uk/news/world/europe/eu-admits-it-has-no-power-to-enforce-its-binding-2030-renewable-energy-targets-9078390.html>.

174 European Council, *Conclusions on 2030 Climate and Energy Policy Framework*, SN 79/14 (23–4 October 2014).

175 Bawden, above n 175.

176 Renewable Energy Target Scheme, *Report of the Expert Panel* (August 2014) <www.aph.gov.au/DocumentStore.ashx?id=17008e4b-e2f3-4ea3-9d53-fb3fb0cd4d85&subId=351098 >.

177 Ibid.

recommended that the revised 20 per cent target for large-scale generation should be achieved through a series of yearly targets set one year in advance that correspond to 50 per cent of growth in electricity demand.[178] On 23 June 2015, these changes were adopted through legislative amendments to the existing RET scheme.[179]

Where RETs are binding on a country, they have proven to be highly effective at encouraging accelerated deployment of renewable energy. This is particularly the case where the RETs have set technology-specific targets as a signal to investors and project developers. However, as shown above in Chapter 4, countries often seek to achieve multiple, and at times conflicting, objectives through their renewable energy laws. In this environment, for a RET to be successful, first the baseline position must be understood, and second, the target must be appropriate for the national political, economic, social and institutional context.[180]

5.2.7 *Subsidies*

Subsidies have been defined by the WTO as a financial contribution made by a government or any other public body within the territory of a member country that confers a benefit.[181] Subsidies have been used within the energy sector since the mid-1800s, when the United Kingdom Government first provided subsidies for the use of coal in order to accelerate the Industrial Revolution.[182] Since then, subsidies have been used to support a range of energy transitions, from supporting the development of oil and natural gas production and use, to more recently aiding the development of the nuclear energy industry. Many countries continue to provide subsidies either for the production or use of conventional fossil fuels. The provision of these subsidies is problematic because subsidising energy prices distorts the market signals that govern supply and demand by masking the true cost of fossil fuel generation and leading to higher use. Ironically, it is the long-term and ongoing provision of these subsidies to conventional fossil fuels that has created a significant market barrier to the development of the renewable energy sector.

178 Ibid.

179 *Renewable Energy (Electricity) Amendment Act 2015* (Cth).

180 See e.g. IRENA, *Renewable Energy Target Setting* (IRENA, Abu Dhabi, June 2015) 9.

181 *Agreement on Subsidies and Countervailing Measures*, WTO Doc 1869 UNTS 14 (15 April 1994) Arts 1, 1.1.

182 Jonathan Pershing and Jim Mackenzie, 'Removing Subsidies: Levelling the Playing Field for Renewable Energy Technologies' (Paper presented at the International Conference for Renewable Energies, Bonn, 2004) 1.

Indeed, one of the strongest arguments supporting the provision of subsidies to the renewable energy sector is to enable competition with existing subsidised conventional fossil fuel sources. There have been recent efforts by the OECD, World Bank, IEA[183] and the Group of Twenty (G20)[184] to advocate for the reduction and eventual removal of subsidies to the conventional fossil fuel sector. However, for as long as these subsidies exist, they will provide a rationale for the provision of subsidies to the renewable energy sector to enable them to compete.

Subsidies to the renewable energy sector take two main forms: investment subsidies and consumer subsidies. Investment subsidies may take a number of forms including rebates, clean energy loans, investment tax credits and production tax credits. Each of these is discussed in more detail below. Where these subsidies solely target investment, rather than production, they are seen as problematic as there is no incentive to operate the renewable energy project efficiently, innovate or lower costs. Consumer subsidies may take the form of direct payments, reduced prices, or low-interest loans to incentivise consumers to install or use renewable energy. Examples of these schemes include the National Solar Schools Program in Australia, which provided schools with a grant of up to $AUD50,000 to support the installation of solar panels,[185] and Finland providing up to 40 per cent of eligible project costs for renewable energy projects with a fixed asset investment above €5,000,000.[186]

Pershing and McKenzie have argued that subsidies for renewable energy should be directed towards: '(1) reducing technical barriers; (2) overcoming market impediments (including through internalising externalities); and (3) addressing administrative barriers and social and environmental constraints'.[187] However, due to the market-distorting effect of subsidies and their potential impact upon the government's budget, a number of considerations need to be taken into account prior to establishing and implementing a renewable energy subsidy programme. First, all existing subsidies directed at

[183] IEA, OPEC, OECD and World Bank, 'Joint report by IEA, OPEC, OECD and World Bank on fossil-fuel and other energy subsidies: An update of the G20 Pittsburgh and Toronto Commitments' (Paper prepared for the G20 Meeting of Finance Ministers and Central Bank Governors and the G20 Summit, France, October–November 2011).

[184] See e.g, the G20 Pittsburgh and Toronto Commitments.

[185] Energy Matters, *Australian Solar Schools Program* (2012) <www.energymatters.com.au/renewable-energy/solar-power/grid-connected-systems/solar-for-schools.php>.

[186] Valtioneuvoston asetus uusiutuvan energian ja uuden energiateknologian investointituen myöntämisen yleisistä ehdoista 25.02.2016/145 [Government Decree No. 145/2016 on Granting Investment Aid for Renewable Energy and New Energy Technologies] (Finland), §5, §10.

[187] Pershing et al., above n 185, 15.

conventional fossil fuel generation and use should be decreased and eventually discontinued. The removal of conventional fossil fuel subsidies should prevent over-expenditure on energy subsidies, because renewable energy will no longer need to be subsidised to compete with subsidised conventional fossil fuels. Second, a cost-benefit analysis should be carried out prior to the introduction of a subsidy to ensure that the subsidy is warranted and is set at an appropriate level. Third, if renewable energy is to be subsidised, the subsidy should be appropriately targeted, transparent, provided for a limited time span and provided through 'competitive mechanisms to ensure that excess "rents" are dissipated'.[188] The requirement that subsidies are available only for a limited lifespan is particularly important. The provision of long-term and ongoing subsidies, particularly when they are linked to investment rather than performance, stifles innovation and competition within the sector. A further concern regarding the lifespan of energy subsidies is that it is often politically difficult to discontinue them. This is because of the immediate short-term economic impacts that the removal of energy subsidies would have on energy prices. Any sudden movements in energy prices will have particular impacts on the cost of living of those on low incomes and, as a result, will be politically unpopular. A fourth concern that needs to be considered prior to the introduction of a subsidy programme (if the country introducing it is a member of the WTO), is whether the programme will comply with the Agreement on Subsidies and Countervailing Measures.[189] As will be shown in Chapter 6, there have been a number of referrals made to the WTO in respect of subsidies that have been designed, and then subsequently implemented, in contravention of these rules. This behaviour is particularly prevalent among countries that are trying to establish a foothold in the renewable energy market, as a common tactic is the inclusion of domestic content clauses in their subsidies in order to bolster their domestic renewable energy technology manufacturing sector. Therefore, while the provision of subsidies can be an effective tool to accelerate the deployment of renewable energy, careful consideration must be given to avoiding the negative impacts associated with their use.

[188] Ibid 14.

[189] *Marrakesh Agreement Establishing the World Trade Organization*, opened for signature 15 April 1994, 1867 UNTS 3 (entered into force 1 January 1995) annex 1A ('Subsidies and Countervailing Measures'); Yulia Selivanova, 'The WTO Agreements and Energy' in Kim Talus (ed.), *Research Handbook on International Energy Law* (Edward Elgar, 2014) 275, 302–5; Anton Ming-Zhi Gao, 'Promotion of Renewable Electricity: Free Trade and Domestic Industrial Development' in Kim Talus (ed.), *Research Handbook on International Energy Law* (Edward Elgar, 2014) 407.

5.2.8 *Clean Energy Loans*

As discussed in Chapter 3, one of the major barriers to the widespread deployment of renewable energy is the high initial capital costs involved with establishing new renewable energy projects. As a result, the cost and availability of debt and project financing have a major impact on the long-term viability of the renewable energy sector.[190]

One of the difficulties in seeking financing for renewable energy projects from private lenders is the risk profile associated with new and emerging renewable energy technologies. As these technologies are often unproven on a large scale, they may not be 'bankable', particularly if project financing is required. This is because the lender needs to ensure that the project will generate sufficient revenue to ensure that the loan will be repaid. However, even in circumstances where private lenders are willing to lend to renewable energy projects, the loans are likely to attract high interest rates and shorter loan terms, reflecting the perceived risk of lending to projects using new technologies.[191] This can add significant costs to renewable energy projects.

In an attempt to address these challenges, ninety-nine countries use clean energy loans as one of their secondary regulatory support mechanisms to promote the accelerated deployment of renewable energy.[192] To establish a clean energy loan programme, governments provide the initial loan pool, which then operates as a revolving loan facility. The Clean Energy Finance Corporation of Australia, which has its investment pool funded by grants from consolidated revenue totalling $AU10 billion, is an example of a clean energy loan programme.[193] Some international institutions such as the World Bank or the Asian Development Bank also provide clean energy loans to developing nations.

Clean energy loan programmes may be either 'administered directly by a government agency or through a public-private partnership in which the program is administered by a private financial institution'.[194] Common features of clean energy loan programmes are that they offer:

- lower interest rates than would be available on the private lending market;

[190] Sawin, above n 3, 20; CESA, *Developing an Effective State Clean Energy Program*, above n 130, 1.
[191] Ibid.
[192] REN21 Secretariat, 'Renewables 2018 Global Status Report', above n 38, 64–7.
[193] Clean Energy Finance Corporation, *Annual Report of the Clean Energy Finance Corporation 2016–7* (CEFC of Australia, 2013).
[194] CESA, *Developing an Effective State Clean Energy Program*, above n 130, 1.

- longer amortisation periods, with repayment terms often reflecting a conservative estimate of the anticipated life of the project (that is, ten years or more);
- simplified application and administrative processes, especially for smaller renewable energy projects; and
- loans without a debt service coverage requirement and without additional secured charges over property that is not the subject of the loan.[195]

The loans granted under a clean energy loan programme vary considerably but common types include: direct loans,[196] matching loans[197] and interest rate buydowns.[198]

Clean energy loan programmes have a number of benefits for governments. First, it provides the government with certainty as to the cost of the programme over time, as the value of the loan is known upfront and, just as with private lenders, the default rate on the loans can be predicted and minimised with proper evaluation and monitoring processes.[199] Second, clean energy programmes are relatively easy to administer[200] and, if run successfully, may actually generate substantial private sector investment, which in turn will increase acceptance amongst other lenders in regard to the financing of renewable energy projects.

Malaysia's Green Technology Financing Scheme is an example of a successful scheme which has provided preferential loans worth $US855 million borrowed for up to fifteen years at interest rates two basis points lower than commercial market rates. In addition, this scheme provides a credit guarantee for 60 per cent of the loan and other capacity building support.[201] Between 2010 and 2017, the scheme supported 272 projects operated by Malaysian companies. It is said to have led to market stimulation of twice the value of the loans, built partnerships with twenty-six financial institutions,

195 Ibid; Ellingson, above n 76, 44.

196 Direct loans are where the government acts as both loan underwriter and servicer: CESA, *Developing an Effective State Clean Energy Program*, above n 130, 1–2.

197 Matching loans are where the government provides a share of the total figure to be borrowed at below market rates on the condition that the borrower must find a commercial lender to provide the balance of the loan amount: Ibid 1–2.

198 Interest rate buydowns are where the government either (a) 'subsidises the interest rate offered by a private lender for a qualified loan'; or (b) 'provides a lump sum payment to the lender in exchange for the lender offering a below-market interest rate': Ibid 2.

199 Mir-Artigues et al., above n 5; Ellingson, above n 76, 44.

200 Mir-Artigues et al., above n 5.

201 Sopitsuda Tongsopit et al., Designing Renewable Energy Incentives and Auctions: Lessons for ASEAN, USAID Clean Power Asia, 4 September 2017, 119–28.

and created 4,000 jobs.[202] Indeed, the Malaysian Government viewed the scheme as so successful that it has extended it for a further five-year term until 2022.[203]

Indeed, the only downside to adopting a clean energy loan programme as one of the country's secondary regulatory support mechanisms is the need for the country to be sufficiently wealthy to be able to dedicate the funds to establish the initial pool of capital from consolidated revenue. This may be difficult to do in lower income countries, which may have more pressing problems such as healthcare or education that need to be addressed first. It is likely that this explains why almost two-thirds of the countries with clean energy loan programmes are high income countries or upper-middle income countries.[204]

5.2.9 *Rebates*

Rebates are lump-sum payments paid to the owner of a renewable energy project to cover a portion of the initial capital cost of that project. They are designed to provide 'a temporary incentive to encourage investment until such time as prices decline to the point of becoming cost competitive in the marketplace'.[205] As such, rebate programmes focus on reducing the high capital costs associated with purchasing and installing renewable energy projects, while increasing consumer awareness and creating a demand in the market for the new technology.

The proportion of the costs to be covered by the rebate is usually determined after an examination of the available funds, desired size of the market, the cost of alternatives and a study of existing market trends.[206] Ellingson has stated that usually rebate programmes seek to cover 20 to 50 per cent of total project costs.[207] However, Sawin, Haas et al. and the Clean Energy States Alliance have criticised the approach of providing a percentage of the total investment in the project, stating that this does not encourage investors to seek out the most cost-effective and efficient option.[208] Instead, they advocate

202 Ibid 126.

203 Ibid.

204 REN21 Secretariat, 'Renewables 2018 Global Status Report', above n 38, 64–7.

205 CESA, *Developing an Effective State Clean Energy Program*, above n 130, 1.

206 Ellingson, above n 76, 53; CESA, *Developing an Effective State Clean Energy Program*, above n 130, 1.

207 Ellingson, above n 76, 52.

208 Sawin, above n 3, 20; Haas et al., 'How to Promote Renewable Energy Systems Successfully and Effectively', above n 8; CESA, *Developing an Effective State Clean Energy Program*, above n 130, 1.

providing a fixed dollar amount per watt of installed generating capacity.[209] These amounts can also be capped to provide a maximum amount per project or banded to provide different levels of support depending on project size.

Some countries provide rebates as an upfront payment to purchasers of renewable energy technologies.[210] Others only provide rebates to the project owner upon project completion.[211] Where the rebate is paid upfront, rebate programmes can be particularly effective during periods of high interest rates and limited capital availability because they reduce the total amount of funds that need to be borrowed.[212] Rebates also have the additional advantage of providing the same benefit to all project developers regardless of their income. Sawin has argued that this not only means that rebates are more equitable than tax credits but ensures that they also result in smoother growth over time rather than encouraging people to invest towards the end of a tax year.[213]

There are a number of downsides associated with the use of rebate programmes. Unlike tax credits, rebate programmes require direct funding from central government. They may, therefore, be subject to instability and uncertainty if the budget levels are frequently altered. In addition, where the level of the rebate granted is not linked to performance or generation of electricity, rebates can distort the market without adequate cost recovery. The use of rebates that provide a proportion of the investment costs rather than being linked to the level of performance also means that projects may be located in less favourable locations or use technologies which do not provide efficient levels of performance.[214] To counter these problems, rebates should be designed as performance-based incentives, linked to the measured output of the renewable energy project over a specified period. In addition, governments should ensure the long-term continuity and stability of the project by guaranteeing funding for five to ten years, with the rebate levels gradually declining to reflect the declining costs of the technology.[215]

5.2.10 *Tax Incentives*

A wide range of tax incentives and concessions are available to participants in the sector to improve the competitiveness of investing in renewable energy.

209 Ibid.
210 CESA, *Developing an Effective State Clean Energy Program*, above n 130, 1.
211 Ellingson, above n 76, 52.
212 CESA, *Developing an Effective State Clean Energy Program*, above n 130, 1.
213 Sawin, above n 3, 20.
214 Ellingson, above n 76, 52–3; CESA, *Developing an Effective State Clean Energy Program*, above n 130, 2.
215 Ibid.

Indeed, tax incentives are used in 108 countries to promote renewable energy.[216] Countries use a range of investment tax credits and production tax credits, as well as tax deductions and exemptions,[217] to provide:

- Full or partial relief from income or corporate tax for renewable electricity. Income tax relief is directly available in France, with deductions available at the rate of 30 per cent of the cost of PV solar equipment installed at a taxpayer's principal residence.[218] Income tax relief is also indirectly available in some countries through favourable depreciation rules and enhanced capital allowances[219] or, in the case of Greece, stabilisation of the income tax coefficient.[220]
- Exemptions for qualifying renewable energy generators from energy, pollution or carbon taxes such as in the Ukraine and United Kingdom.
- Lower rates of value added tax (VAT) applied to sales of qualifying renewable energy technologies such as in France, Indonesia and Italy.[221]
- Property tax reductions for land used to locate renewable energy projects such as in Italy.[222]
- Many countries also provide a 'reduction or elimination of import duties for renewable energy technologies or components'[223] to reduce the high initial capital costs of renewable energy projects. This is especially valuable in countries without a strong domestic manufacturing industry.[224]

216 REN21 Secretariat, 'Renewables 2018 Global Status Report', above n 38, 64–7.

217 Tax deductions permit either the full or a partial amount of qualifying expenses to reduce the gross amount of tax owed. Alternatively, they may operate to exclude the operation of, or reduce the level of, sales taxes, VAT, energy or carbon taxes from eligible projects.

218 *Code Général des Impôsts, version consolidée au 1 septembre 2018* [General Tax Code, version consolidated to 1 September 2018] [Legifrance (Government of France) translation from French], Art 200, quarter par. 1c, 5.

219 Kitzing et al., above n 44, 195; Van der Linden et al., above n 33, 12.

220 Νόμος 4399/2016 – Νόμος 4399/2016 Θεσμικό πλαίσιο για τη σύσταση καθεστώτων Ενισχύσεων Ιδιωτικών Επενδύσεων για την περιφερειακή και οικονομική ανάπτυξη της χώρας – Σύσταση Αναπτυξιακού Συμβουλίου και άλλες διατάξεις [Law 4399/2016 Institutional Framework for the Establishment of Private Investment Aid Schemes for the Regional and Economic Development of the Country – Establishment of a Development Council and Other Provisions] (Greece) [Start-up Greece (Government of Greece) translation from Greek], Art 66.

221 *Code Général des Impôsts, version consolidée au 1 septembre 2018* [General Tax Code, version consolidated to 1 September 2018] [Legifrance (Government of France) translation from French], Art 279; Res Legal, *Compare Support Schemes* (2018) <www.res-legal.eu/en/compare-support-schemes>; Sopitsuda Tongsopit et al., above n 204, 111.

222 Res Legal, *Compare Support Schemes* (2018) <www.res-legal.eu/en/compare-support-schemes>.

223 Sawin, above n 3, 19.

224 Ibid.

Kitzing et al. also note that by enabling consumers to engage in net metering, governments are also providing a degree of tax relief to consumers from the volumetric taxes that are applicable to electricity usage such as energy taxes, carbon taxes and VAT.[225] Another benefit for renewable energy generators is that they are not subject to some of the environmental and carbon taxes to which other conventional fossil fuel electricity generators are subject.

Tax incentives are a popular regulatory support mechanism as they are comparatively simple to administer, effective in lowering the investment and production risks associated with the sector and prima facie do not need to be funded out of consolidated revenue as they merely reduce the amount of tax collected.

5.2.10.1 Investment Tax Credits

Investment tax credits provide a full or partial tax credit for investments in renewable energy technologies. In some cases, the eligibility for these investment tax credits may extend to the installation of these technologies. Investment tax credits work by reducing the high capital costs associated with investing in new renewable energy technologies and therefore reduce some of the risk of investment.[226] They are also beneficial in supporting dynamic efficiency as they can be specifically targeted to support less mature technologies.[227]

Sovacool has stated that despite these benefits, investment tax credits also have a number of disadvantages. He states that the investor still needs to be able to afford to make the high initial upfront payments and then must wait until their tax return has been processed before receiving the credit.[228] Further, because investment tax credits target investment in a technology rather than the comparative performance of the technology, they may send the market incorrect signals as to which technology to invest in.[229] The design of investment tax credits also does not provide an incentive to drive down the costs of renewable energy technologies.[230] Despite this, when coupled with production tax credits, there are clear market signals for selecting to invest in technologies that will, once the technology is installed and generating electricity, be both efficient and cost-effective.

225 Kitzing et al., above n 44, 195.
226 Sawin, above n 3, 18; The Energy and Resources Institute, above n 22, 19.
227 Sovacool, *Renewable Electricity for Southeast Asia*, above n 3, 20.
228 Ibid.
229 Ibid.
230 Ibid.

5.2.10.2 Production Tax Credits

Production tax credits provide the owner or investors in qualifying renewable energy projects with tax credits calculated on the basis of the number of kWh of electricity generated by the project and fed into the grid within the tax year. Production tax credits reward efficient performance and the cost-efficient production and supply of renewable energy into the grid. This is because the cheaper the cost of generating and supplying electricity, the greater the profit from the production tax credit. Indeed, the American Wind Energy Association claims that the US Federal Production Tax Credits were instrumental in spurring investment and driving 'US wind power costs down by 67% in the last seven years'.[231] This far exceeds the predicted benefits from a 2007 study, which estimated that the benefits of extending production tax credits to the wind industry in the United States for ten years would result in predicted cost savings for wind turbines of 22 per cent or $US380 per installed kW over the period.[232] It should be noted that, due to their structure, production tax credits generally favour larger renewable energy projects rather than providing investment in smaller-scale projects.[233] However, this problem may be lessened when both production tax and investment tax credits are available to investors and owners, as, depending on the design of the investment tax credit, they are likely to fulfil the role of supporting small-scale projects.

5.2.11 *Public Benefit Funds*

Public benefit funds, also known as 'systems benefits funds' or 'clean energy funds', involve charging customers a small tax per kWh of electricity used. In some countries, instead of being linked to electricity consumption, a small fixed fee is charged instead to customers as part of their electricity bill.[234] The funds collected under these programmes are then used by countries to pursue a range of socially beneficial energy projects, such as removing the technical, regulatory and market barriers to emerging renewable technologies.[235] These funds tend to be used in three different ways to:

231 American Wind Energy Association, *Tax Policy* (September 2018) <www.awea.org/policy-and-issues/tax-policy>.

232 Ryan Wiser, Mark Bollinger and Galen Barbose, 'Using the Federal Production Tax Credit to Build a Durable Market for Wind Power in the United States' (2007) 20(9) *The Electricity Journal* 77, 84.

233 Sovacool, *Renewable Electricity for Southeast Asia*, above n 3, 21.

234 Ellingson, above n 76, 23–4.

235 Ibid.

1. target investment in renewable energy programmes through the provision of loans;
2. promote project development through the provision of competitive grants, rebates and production incentives; or
3. support the development of the renewable energy industry by aiding research and development, providing technical assistance, consumer education and financing demonstration projects.[236]

Most countries use their public benefit funds to support a diverse portfolio of programmes and incentives.

Public benefit funds appear to be most commonly found in some of the American states. They are thought to be fiscally responsible, as the funds are derived directly from ratepayers rather than through consolidated revenue.[237] Sovacool has also argued that they promote dynamic efficiency because they enable policymakers to support a broad array of technologies and projects, including those that may benefit low-income consumers.[238] Despite this, questions have been raised as to whether public benefit funds promote efficacy, cost-effectiveness and equity.[239] This is because, while the funds may be used to support projects that benefit low-income consumers, the beneficiaries of these funds tend to be corporations or foundations that have the skills and capacity to put together proposals to access the funds.[240] Further, because public benefit funds tend to provide lump sum support, there is little incentive to innovate over time.

5.2.12 *Research and Development Support*

It is often said that there are five stages to technological innovation: research and development; demonstration; deployment into niche markets; diffusion; and commercial maturity. The first two stages of technological innovation are highly capital-intensive, with significant risks of failure. This, coupled with the current cost differential between the renewable energy and fossil fuels (which have artificially low prices due to the subsidies given to conventional fossil fuels and the failure to internalise the costs of externalities), provides a justification for the support of research and development.

236 Ibid.
237 Sovacool, *Renewable Electricity for Southeast Asia*, above n 3, 19–20.
238 Ibid.
239 Ibid.
240 Ibid.

Research and development support is a supply-side mechanism designed to encourage the development of new renewable energy technologies and reduce the cost of existing technologies through technological improvements. This mechanism is quite commonly adopted as a secondary instrument. All ten jurisdictions[241] investigated by Fischer and Preonas in their study of policies promoting renewable energy in the electricity sector reported that they used it as a support mechanism.[242] This research also shows that research and development support, as a secondary instrument, is often used in combination with feed-in tariffs (eight out of ten jurisdictions) or renewable portfolio standards/quota obligations (five out of ten jurisdictions).[243] It also often co-exists with other secondary instruments such as subsidies (three out of ten jurisdictions) and tax incentives (five out of ten jurisdictions).[244]

The bulk of the existing research and development support provided to the renewable energy sector is provided by governments, though private companies are also involved in providing this support, both on an individual company basis and through industry-wide partnerships with research institutions. Governments also have the ability to mandate that private companies dedicate a specified portion of their profits towards research and development. This mandatory approach for private companies is not often used, with governments preferring to offer substantial tax breaks to encourage voluntary participation instead.

Ottinger et al. have noted that in recent years, 'corporations have significantly decreased their long-term research and development expenditures. Governments have done the same thing from budgetary concerns'.[245] The impact of these reductions in expenditures is heightened because countries already diverge widely in their level of commitment depending on whether they view the renewable energy sector as critical to their industrial policy and economic development, or are technology laggards. Grafström et al. have found that there is a significant divergence of practice even within the EU with 'innovation activities in the renewable energy sector [...] typically concentrated [in] a few leading economies', while other Member States engage in free-riding behaviour.[246] Further, Mitchell has expressed a concern that

241 The jurisdictions considered in the study were Canada, Denmark, Germany, Japan, The Netherlands, New Zealand, Norway, Spain, the United Kingdom, the United States of America (Federal) and the United States of America (states).

242 Fischer and Preonas, above n 7, 60.

243 Ibid.

244 Ibid.

245 Ottinger et al., above n 2, 195.

246 Grafström et al., 'Knowledge Accumulation from Public Renewable Energy R&D in the European Union: Converging or Diverging Trends?' (UFZ Discussion Papers, Department

despite research and development support being provided in the United Kingdom, it is yet to materially change the nation's energy outlook.[247] She attributes this 'in part to the fact that similar funding to nuclear and fossil fuels has dwarfed that for renewables, but it also can be taken as a general indictment of research funding as a renewables promotion device'.[248] This is because even when adequate research and development support is provided, sufficient market demand must exist to support the commercialisation of technologies developed using this supply-side mechanism.[249] Sovacool has further expressed concerns that, while research and development support promotes dynamic efficiency by being flexible and enabling a wide range of projects and applications, it may not necessarily be efficient, cost-effective, equitable or fiscally responsible.[250] Many of these criticisms seem to be attributable to the fact that research and development support relies upon substantial government support and hence is subject to budgetary pressures.[251] Further, the projects selected for support may not necessarily be the most cost-effective, and will, by necessity, tend to be located at bigger institutions or corporations that have the specialist equipment and support to conduct this research.[252]

5.2.13 *Green Power Schemes*

Green power schemes or 'green marketing' are voluntary programmes, which provide consumers with the option to purchase all or a portion of their electricity from guaranteed sources of 'green' power for a premium.[253] This premium often takes the form of consumers paying a higher rate for each kWh of green power consumed, reflecting the higher costs involved in generating green power. Green power schemes are used in a number of predominately higher income countries. However, due to the fact that they are run voluntarily by electricity supply companies, precise figures on the number of countries in which they are presently offered are not available.

of Economics 5/2017, Helmholtz-Zentrum für Umweltforschung GmbH-UFZ, September 2017) 4, 18.

247 Mitchell, *The Political Economy of Sustainable Energy*, above n 4, 51.

248 Ibid.

249 Ibid.

250 Sovacool, *Renewable Electricity for Southeast Asia*, above n 3, 18.

251 Ibid 19.

252 Ibid.

253 See generally Rosemary Lyster and Adrian Bradbrook, *Energy Law and the Environment* (Cambridge University Press, 2006) 155–63.

Two key factors affect the successful implementation of green power schemes. First, as these are voluntary programmes, without the development of an industry standard, the definition of green power may vary significantly between different electricity supply companies. In particular, where the term 'alternative energy sources' has been used in renewable energy laws, often it does not just include renewable energy but may also include other sources of energy such as 'clean coal' (which is electricity produced using low greenhouse gas emitting supercritical coal-fired generators).[254] Therefore, it is important that the definition of green power used in green power schemes is clearly defined and reflects common understandings of that phrase: that is, only including renewable energy sources. This approach has been adopted in the United States, where the electricity industry uses the definition from the United States Environmental Protection Agency (US EPA) as its market standard. The US EPA defines green power as 'electricity generated or used from renewable energy sources with low or no environmental impacts … [that] goes above and beyond what is otherwise required by mandate or requirement – it is voluntary or surplus to regulation'.[255] They further stipulate that due to their environmental impacts, large scale hydropower and municipal solid waste are generally not deemed to be green power despite being defined as renewable energy sources.

A second variable that will affect the ultimate success of any green energy programme is the willingness of electricity customers to pay a higher price for renewable electricity. Willingness to pay varies considerably by country, and is related to consumer awareness of environmental issues and specific market conditions. The level of price differential from electricity generated using conventional fossil fuel sources is also a relevant factor, leading to green power schemes generally supporting generation from the most commercialised sources of renewable energy in order to keep green power prices low.[256] The US EPA has reported that between 2006 and 2015, the average tariff for green power products for a residential consumer was $US0.02 per kWh above the standard electricity tariff.[257] For the average American household, this equates to an annual premium of $US216.[258] Ottinger et al. have reported that in most

254 *Law on Alternative Energy Sources* (Ukraine) 20 February 2003, No. 555-IV, Art 1 [Linguistico Translations translation from Ukrainian].

255 United States EPA *Green Power Partnership Glossary* (21 December 2017) <www.epa.gov/greenpower/green-power-partnership-glossary>.

256 Sovacool, *Renewable Electricity for Southeast Asia*, above n 3, 17.

257 United States EPA, *Green Power Pricing* (21 December 2017) <www.epa.gov/greenpower/green-power-pricing>.

258 Ibid.

countries with green power schemes, approximately 1 per cent of electricity consumers are willing to pay the higher prices involved in joining a green power scheme,[259] while Sovacool has stated that participation rates rarely exceed 5 per cent.[260] In 2016, 3.21 per cent of electricity in the United States was purchased under green power schemes.[261] This may change over time, with the Netherlands (whose population express higher levels of environmental awareness and concern) displaying uptake rates of approximately 13 per cent of all electricity customers.[262] Van der Linden has theorised that the low uptake levels in many countries may at least in part be due to 'consumer scepticism about the premium being used effectively to promote renewables'.[263]

With these low uptake levels and due to their voluntary nature, green power schemes cannot be relied upon as a primary mechanism for accelerating the deployment of renewable energy. Further, there is a tendency for the majority of consumers to be 'free-riders' and to not change their consumer behaviour.[264] Despite this, these schemes are seen as valuable because they provide customers with a choice of purchasing electricity generated from renewable sources, increase customer awareness of the availability of electricity generated from renewable sources and create acceptance for other regulatory support mechanisms.[265]

5.2.14 *Other Strategies*

In addition to the regulatory mechanisms that are often found in the primary renewable energy legislation detailed above, a number of other strategies exist to promote the accelerated deployment of renewable energy. These include:

1. internalising the externality costs associated with conventional fossil fuels through the introduction of carbon taxes, ETS and other pollution pricing mechanisms;[266]
2. providing improved transmission planning using anticipatory transmission planning processes. Most transmission planning occurs on

259 Ottinger et al., above n 2, 198.

260 Sovacool, *Renewable Electricity for Southeast Asia*, above n 3, 16.

261 Office of Air, United States Environment Protection Agency, *Guide to Purchasing Green Power: Renewable Electricity, Renewable Energy Certificates, and On-Site Renewable Generation* (September 2018) vii <www.epa.gov/greenpower/guide-purchasing-green-power>.

262 Ottinger et al., above n 2, 198.

263 Van der Linden et al., above n 34, 12.

264 Bradbrook, above n 3, 23–4; Sovacool, *Renewable Electricity for Southeast Asia*, above n 3, 17.

265 Bradbrook, above n 3, 23.

266 Ottinger et al., above n 2, 202–3.

a reactive basis, which means that planning, reinforcement and/or construction of transmission and distribution lines does not occur until after a renewable energy project developer has made a request for the transmission interconnection and service. Reactive planning can add considerable uncertainty and create delays for renewable energy projects. For example, the planning process for the Beauly-Denny transmission line in Scotland, which was urgently required to enable new renewable generation projects access to the transmission and distribution network, spent over six years under consideration by the authorities before approval was finally granted.[267] Moving to anticipatory transmission processes, where transmission planning and, in some cases, construction, occurs prior to a formal request from a renewable energy project developer will lessen uncertainty by providing project developers greater clarity about how, when and where transmission access and interconnection are likely to be granted;[268]

3. encouraging the use of renewable energy in government procurement to foster demand for renewable energy.[269] The US EPA has recommended that state and local governments should consider aggregated purchasing of renewable energy, so that government agencies do not need individually to negotiate power purchase agreements and to enable access to bulk purchase discounts;[270]
4. providing education and training is another common strategy used by countries to promote the accelerated deployment of electricity generated from renewable energy sources. Ottinger et al. have argued that the general public, energy decision-makers and the private sector need to be educated about 'the external costs of fossil fuels, the need to reduce carbon dioxide emissions, and the available renewable energy options, applications, costs, and benefits'.[271] Many countries such as Australia[272]

267 Kristy Dorsey, 'Beauly-Denny: Shock to the System', *Scotland on Sunday (Edinburgh)* (online), 9 January 2010 <www.scotsman.com/business/beauly-denny-shock-to-the-system-1-1363889>.

268 Alagappan et al., above n 124, 5101–3.

269 Ottinger et al., above n 2, 200; Ellingson, above n 76, 96–7.

270 United States Environmental Protection Agency, *Clean Energy Lead by Example Guide* (2009) 10 <www.epa.gov/sites/production/files/2015-08/documents/state_lead_by_example_guide_full_report.pdf>.

271 Ottinger et al., above n 2, 200.

272 New South Wales Department of Education and Communities, *Sustainability: Learning Across the Curriculum* (2018) New South Wales Government <https://education.nsw.gov.au/teaching-and-learning/curriculum/learning-across-the-curriculum/sustainability>.

and Germany[273] now include energy issues, including those associated with renewable energy generation, as a core component of the school curriculum. Meanwhile, the Ministry of New and Renewable Energy of the Indian Government runs a number of programmes utilising electronic and print media, radio advertising, exhibitions and outdoor advertising to disseminate information on renewable energy and promote its uptake;[274]

5. in developing countries, a number of non-governmental organisations (NGOs) have partnered with private sector enterprises to overcome the inability of governments in low-income countries to fund large clean energy loan programmes by introducing micro-finance and leasing schemes to support small-scale renewable energy projects.[275] These schemes enable consumers either to purchase outright or lease small renewable energy systems (thereby removing the need for consumers to bear the high upfront capital equipment costs). The micro-finance loans issued under these schemes are often aggregated, with banks lending to a local community association to avoid the costs associated with servicing many small loans.[276] One of the other features of these micro-finance loans is that they can be tailored to reflect local social and economic conditions. For example, a key feature of a programme that saw 140,000 small-scale wind turbines installed in Inner Mongolia and successfully producing power for more than 500,000 people was that the loan repayments were scheduled to coincide with harvest season and the future sales of cattle or wool.[277]

The use of these policy-based strategies in combination with the regulatory support mechanisms outlined above to target particular market failures and

[273] Gerhard De Haan, 'The BLK "21" Programme in Germany: A "Gestaltungskompetenz"-based Model for Education for Sustainable Development' (2006) 12 *Environmental Education Research* 19.

[274] Ministry of New and Renewable Energy, *Support programmes* (2014) Government of India <https://mnre.gov.in/support-programmes>.

[275] United Nations Development Programme and United Nations Capital Development Fund, *Clean Start: Microfinance opportunities for a clean energy future* (2013) UNCDF <www.uncdf.org/sites/default/files/Documents/cleanstart_publication.pdf>.

[276] See e.g. Ibid; P Sharath Chandra Rao, Jeffrey B Miller, Young Doo Wang and John B Byrne, 'Energy-microfinance Intervention for Below Poverty Line Households in India' (2009) 37 *Energy Policy* 1694; Kadra Branker, Emily Shackles and Joshua M Pearce, 'Peer-to-peer Financing Mechanisms to Accelerate Renewable Energy Deployment' (2011)1 *Journal of Sustainable Finance & Investment* 138.

[277] Eric Martinot et al., 'Renewable Energy Markets in Developing Countries' (2002) 27 *Annual Review of Energy and the Environment* 309, 318–19.

barriers provide a number of advantages. First, policies are often more flexible than regulatory support mechanisms and thus can be amended quickly in the event of sudden market shifts. Second, they can be more easily designed to target particular communities or geographic regions as they do not require the same level of political negotiations as legislation. For these reasons, though, policies are often considered to be less effective than legislation in providing stability and certainty as to the government intervention in the sector, as well as sometimes lacking in public legitimacy. Further, as with the regulatory support mechanisms, these interventions also impose costs and their impacts need to be closely evaluated and understood, particularly when they are used in combination with a number of regulatory support mechanisms.

5.3 EVALUATING THE SUCCESS OF REGULATORY SUPPORT MECHANISMS

There is strong evidence that the primary factor encouraging accelerated deployment of renewable energy is the presence of national level regulatory and policy intervention.[278] Given the range of regulatory support mechanism options available to countries seeking to accelerate the deployment of renewable energy, evaluating their relative success or failure within specific national contexts is an important task. As shown above, each regulatory support mechanism presents its own advantages and disadvantages. One of the greatest challenges for governments is deciding which regulatory support mechanisms are most appropriate for their national and local conditions, such that the regulatory support will garner sufficient public support. This process of evaluation ensures that the regulatory support mechanisms adopted within a country meet national and local needs,[279] are cost effective, have static and dynamic efficiency and are equitable.

Numerous criteria have been proposed in the academic literature against which regulatory support mechanisms should be evaluated. However, the approach that seems to have garnered the most support is a test based on the:

- *efficacy* of the mechanism in achieving its objectives in accelerating installed capacity or generation;
- *efficiency* of the mechanism relative to other alternatives (incorporating both static and dynamic efficiency);

[278] See e.g. Sanya Carley et al.., 'Global Expansion of Renewable Energy Generation: An Analysis of Policy Instruments' (2017) 68 *Environmental Resource Economics* 397, 399 and 438.

[279] Ottinger et al., above n 2, 205.

- *equity* of the mechanism in terms of who is paying for the mechanism and who is benefiting; and
- *institutional feasibility*, which considers whether the mechanism is transparent, predictable and likely to be accepted by the industry and the general public.[280]

When regulatory support mechanisms were first introduced to the sector, most countries made a choice between adopting a feed-in tariff (a price-based mechanism) and a renewable portfolio standard/quota obligations (a quantity-based mechanism). Nicolini and Tavoni have found that in Europe these policies were effective in promoting renewable energy, and led to increased production of incentivised energy in the short-term and, in the longer-term, greater installed capacity.[281] Others have noted the numerous empirical studies that have shown that regulatory support mechanisms, particularly FITs, have placed downwards pressure on electricity prices due to the merit-order effect, while positively impacting on innovation, competitiveness and employment.[282] In the past few years, there has been a noticeable shift away from using renewable portfolio standards and, more recently, a few countries have begun to remove their feed-in tariffs. In their place, many countries have implemented competitive tendering due to their ability to achieve deployment at lowest cost.[283]

5.3.1 *Combining Regulatory Support Mechanisms*

The general consensus from much of the research in this field is that there is no single 'best' regulatory support mechanism that will adequately support the development, and subsequent commercialisation, of all renewable energy sources, technologies and scales of renewable energy projects in all countries.[284] The vast majority of countries now use a combination of different regulatory support mechanisms.[285] They will often adopt a primary

280 Catherine Mitchell et al., 'Policy, Financing and Implementation' in Ottmar Edenhofer et al. (eds.), *IPCC Special Report on Renewable Energy Sources and Climate Change Mitigation* (IPCC, 2011); Verbruggen et al., above n 7; IRENA, above n 47.

281 Marcella Nicolini and Massimo Tavoni, 'Are Renewable Energy Subsidies Effective? Evidence from Europe' (2017) 74 *Renewable and Sustainable Energy Reviews* 412.

282 Margarita Ortega-Izquierdo and Pablo del Río, 'Benefits and Costs of Renewable Electricity in Europe' (2016) 61 *Renewable and Sustainable Energy Reviews* 372, 375.

283 Jenny Winkler et al., 'Effectiveness and Efficiency Auctions for Supporting Renewable Electricity – What Can We Learn from Recent Experiences?' (2018) 119 *Renewable Energy* 473; IEA and IRENA, *Perspectives for the Energy Transition: Investment Needs for a Low-Carbon Energy System*' (OECD/IEA and IRENA, 2017) 32.

284 Ottinger et al., above n 2, 206.

285 REN21 Secretariat, 'Renewables 2018 Global Status Report', above n 38, 64–7.

mechanism, commonly competitive tendering, though in some cases still a feed-in tariff, feed-in premium, a renewable portfolio standard or net metering but this will now be supported with a number of secondary mechanisms such as tax incentives, research and development support. In some countries, hybrid mechanisms combining elements of feed-in tariffs, tradeable green certificates and renewable portfolio standards have been developed.[286] Interestingly, the complexity of a country's national support scheme for renewable electricity is strongly correlated to their income level. High-income countries employ an average of 3.6 different policy types, whereas, in low-income countries, it is only 1.5.[287] Not surprisingly, the high-income countries were also much more likely to adopt feed-in tariffs, renewable portfolio standards, tradeable green certificates, net metering, subsidies and soft loans than the low-income countries. In contrast, low-income countries were comparatively more likely to use tendering and tax concessions. This reflects the reality that high-income countries have greater budgets to dedicate to accelerating deployment in the sector, while low-income countries are more likely to select a reduction in their consolidated revenue rather than direct spending.

One of the issues with the growing use of combinations of regulatory support mechanisms is that there is little research available on how different mechanisms interact when used in concert and the impact of this interaction.[288] Much of the previous research has focused on which primary instrument is more efficient in achieving cost effective and technologically diverse deployment. However, this does not consider how these instruments are used in reality by the vast majority of countries. By considering regulatory support mechanisms in isolation, rather than in a regulatory and policy context where different support mechanisms interact with each other, the extent of the interaction and the resultant impact is not known. Research is not only needed where a country has adopted multiple regulatory support mechanisms but also in circumstances where the mechanisms adopted by one country have cross-border implications for the mechanisms of other countries, such as within the European Union. Until these issues are better understood, it is difficult to know which combinations of regulatory support mechanisms might best address the market failures that exist within the sector. In particular, Mir-Artigues and del Río have expressed concerns that 'the

286 Davies, 'Reconciling Renewable Portfolio Standards and Feed-in Tariffs', above n 2, 313; Davies, 'Incentivizing Renewable Energy Deployment', above n 40, 81; Van der Linden et al., above n 34, 61.

287 Author's own calculations from data contained in Table 2, REN21 Secretariat, 'Renewables 2018 Global Status Report', above n 38, 64–7.

288 Mir-Artigues et al., above n 5, 430; Fischer et al., above n 7.

interaction between instruments has been shown to lead to conflicts, resulting in inefficiencies, redundancies, double coverage or double counting'.[289] Similarly, Fischer and Preonas have warned that a lack of coordination between mechanisms will increase the burden on consumers and taxpayers, and possibly lead to the overcompensation of renewable energy generators.[290] Zhao et al. have reported that their research shows that 'as the number of policy instruments increases, especially when the strength of policy instruments reaches a certain level, the policy effects plateau or even decline'.[291] This situation requires close attention where a renewable energy source or technology may benefit from more than one regulatory support mechanism.[292]

5.4 CONCLUSION

Countries have adopted diverse combinations of regulatory support mechanisms to accelerate their deployment of renewable energy. This suggests two conclusions. First, while many countries appreciate that government intervention is required to support the deployment of renewable energy, no one mechanism or combination of mechanisms will meet the needs of every country. Second, different combinations of regulatory support mechanisms may be better suited to meeting the market failures, market barriers and legislative objectives of different countries. Unfortunately, while it is known that there is divergence in the regulatory support mechanisms, the extent of this divergence and the full impact of using a number of mechanisms in combination are unknown. Further research is required to better understand the impact of combining different mechanisms on the operation and development of the renewable energy sector. The use of a combination of regulatory support mechanisms also increases the legislative complexity of the renewable energy sector and may make it more difficult for sectoral-wide soft convergence processes to effectively occur over time.

What is apparent, though, is that the mechanisms adopted in a country need to be tailored to the specific needs of that country in the context of their natural resource endowment, energy market structure, level of development and political and cultural context. Ideally, the mechanisms should set a clear target and be relatively simple to administer, which in turn should reduce the

289 Mir-Artigues et al., above n 5, 431.

290 Fischer et al., above n 7.

291 Yong Zhao, Kam Ki Tang and Li-li Wang, 'Do Renewable Electricity Policies Promote Renewable Electricity Generation? Evidence from Panel Data' (2013) 62 *Energy Policy* 887, 892.

292 Ottinger et al., above n 2, 192.

obligations of monitoring compliance and ensuring enforcement with the mechanism. Reducing complexity in the design and implementation of regulatory support mechanisms will also lower transaction costs and barriers to entry for market participants, making soft convergence easier in the future. Mechanisms that have a long life-span, with the grandfathering of support for existing participants and reductions in the incentive over time (coupled with a clear end date) to reflect learning effects and cost reductions in the technologies, also provide the market with certainty and stability. The mechanism should only be available to new installed capacity, so that existing renewable generation projects cannot benefit from super-profits. In addition, there seems to be consensus that it is appropriate to band the level of support that the mechanism provides in accordance with the level of commercialisation of the technology to ensure that a diverse range of energy sources and technologies are supported. However, the successful design, implementation and enforcement of compliance with mechanisms are not enough; renewable energy projects also require appropriate infrastructure so that they may connect to the transmission and distribution networks. Finally, a clear government commitment over the long term to correct the market failures in the renewable energy sector will decrease the perceived risk within the sector and potentially increase the available capital for investment.

6

The Future Development of Regulatory Support Mechanisms – Unification, Harmonisation, Convergence, Divergence or Regulatory Competition?

The regulation and governance of renewable energy has historically been highly fragmented internationally, with 'no overarching regulation that specifically addresses energy'.[1] There are a number of international organisations that have renewable energy within their purview: the IRENA, the IEA, the Energy Charter Treaty (ECT) and the EU. However, these organisations either have limited membership (the IEA and the EU), or have not actively sought to promote the legal harmonisation of renewable energy law (the ECT) or actively intervened in trade disputes (the IRENA).[2]

From an economic perspective, countries legislate to support the accelerated deployment of electricity generated from renewable energy sources in order to correct a number of market failures that afflict the sector. These market failures were analysed in Chapter 3, and include the failure to price externalities into energy prices, positive spillovers and learning effects and information asymmetries. The existing research on the market failures that afflict the renewable energy sector suggests that the same failures seem to exist in many countries around the world. Unfortunately, to date there has not been a comprehensive analysis of the scale and impact of these market failures in every country that has renewable energy laws. That said, it would be a fair assumption that the market failures affect different countries to varying degrees. The most obvious difference in the market failures is in respect of the failure adequately to price in the positive externalities associated with diversifying supply and ensuring energy security in electricity generated from renewable energy sources.

1 Joanna I Lewis, 'The Rise of Renewable Energy Protectionism: Emerging Trade Conflicts and Implications for Low Carbon Development' (2014) 14(4) *Global Environmental Politics* 10, 28.

2 Ibid. See also Alexandra Wawryk, 'International Energy Law: An Emerging Academic Discipline' in Paul Babie and Paul Leadbeter (eds.), *Law as Change: Engaging with the Life and Scholarship of Adrian Bradbrook* (University of Adelaide Press, 2014) 223, 240.

This market failure will be more of an issue for countries that are energy importers and so have a clear energy security problem, as opposed to the countries that are energy exporters.

However, given that renewable energy laws are largely seeking to address the same underlying problem – that of the presence of market failures affecting the sector – one might assume that similarities may also start to emerge in their legislative solutions. Further, with globalisation contributing to the 'intensification of economic, political, social and cultural relations across borders',[3] the legal and policy convergence literature suggests that over time countries with similar social and economic development should gravitate towards similar policies and instruments.[4]

6.1 WHAT ARE THE ADVANTAGES AND DISADVANTAGES OF NATIONAL RENEWABLE ENERGY LAWS BECOMING MORE SIMILAR INTERNATIONALLY?

There are said to be a number of benefits that are derived from laws becoming more similar across jurisdictions, regardless of whether this is achieved by the processes of legal harmonisation or legal convergence. The presence of similar or identical laws or standards across jurisdictions arguably promotes and facilitates international trade, especially for multinational corporations operating across national boundaries.[5] This is because when companies have to comply with the same laws or standards across jurisdictions, information costs and the barriers to entry to the market are lowered,[6] while legal certainty and

[3] Hans-Henrik Holm and Georg Sørensen, 'Introduction' in Hans-Henrik Holm and Georg Sørensen (eds.), *Whose World Order: Uneven Globalization and the End of the Cold War* (Westview, 1995) quoted in Peter Drahos and John Braithwaite, 'The Globalisation of Regulation' (2001) 9(1) *The Journal of Political Philosophy* 103, 104.

[4] See e.g. Emanuela Carbonara and Francesco Parisi, 'The Paradox of Legal Harmonization' (2007) 132(3/4) *Public Choice* 367, 367–8; Andrew Jordan, Rüdiger Wurzel and Anthony Zito, 'Innovating with "New" Environmental Policy Instruments: Convergence or Divergence in the European Union?' (Paper presented at the 2000 Annual Meeting of the American Political Science Association, Marriott Wardman Park, 31 August–3 September 2000) 6; Daniel W Drezner, 'Globalization, Harmonization and Competition: The Different Pathways to Policy Convergence' (2005) 12 *Journal of European Public Policy* 841, 841.

[5] See e.g. Thomas K Cheng, 'Convergence and Its Discontents: A Reconsideration of the Merits of Convergence of Global Competition Law' (2012) 12 *Chicago Journal of International Law* 433, 455, 461; House of Representatives Standing Committee on Legal and Constitutional Affairs, *Harmonisation of legal systems: Within Australia and Between Australia and New Zealand* (Parliament of Australia, 2006) 9.

[6] Larry E Ribstein and Bruce H Kobayashi, 'An Economic Analysis of Uniform State Laws' (1996) 25(1) *The Journal of Legal Studies* 131, 138.

predictability are improved.[7] The combination of these benefits should in theory lead to increased competition, greater economies of scale and thus lower costs for the end-consumer.[8] Such a move should also lead to reduced compliance costs and minimise cross-jurisdictional conflicts.[9] In addition, when the differences between different countries' laws are reduced, there are also fewer incentives for, and benefits to be derived from, multinational companies engaging in forum shopping.[10] The benefits attached to this are considerable, with Rodrik arguing that the diversity of national institutional arrangements are the most important source of transaction costs in international exchanges, accounting for nearly 35 per cent of the total transaction costs in *ad valorem* terms.[11]

Arguably, one of the greatest benefits of harmonising or converging laws is achieved when international externalities exist in a market, which cannot be resolved by countries operating individually in their own self-interest.[12] For example, while individual countries can act to reduce their own greenhouse gas emissions, addressing climate change requires collective international action to maximise their joint welfare[13] and minimise future negative externalities. However, in order for collective action on international externalities to be successful, countries must have similar objectives for intervening in the market and must be committed to achieving those goals.[14] This is unlikely to be achieved through national renewable energy laws, at least in the short to medium term, as such a consensus is lacking in the renewable energy sector. This is because, as discussed in Chapter 4, 28 different legislative objectives

7 Polina Dlagnekova, 'The Need to Harmonise Trade-related Laws Within Countries of the African Union: An Introduction to the Problems Posed by Legal Divergence' (2009) 15(1) *Fundamina* 1, 23–4.

8 See e.g. Carsten Hefeker, 'The Limits of Economic Policy Convergence in Europe' (2013) 2 *Intereconomics* 83, 85; Dlagnekova, above n 7, 10; Gustav Resch et al., 'Coordination or Harmonisation? Feasible Pathways for a European RES Strategy Beyond 2020' (2013) 24 *Energy and Environment* 147, 158.

9 Cheng, above n 5, 438, 454, 461; Dlagnekova, above n 7, 23.

10 Ribstein et al., above n 6, 138–9; see also Cheng, above n 5, 459.

11 Dani Rodrik, 'Globalization and Growth – Looking in the Wrong Places' (2004) 26 *Journal of Policy Modeling* 513, 514.

12 Aleh Cherp, Jessica Jewell and Andreas Goldthau, 'Governing Global Energy: Systems, Transitions, Complexity' (2011) 2 *Global Policy* 75, 76; Filomena Chirico and Pierre Larouche, 'Convergence and Divergence, in Law and Economics and Comparative Law' in Pierre Larouche and Peter Cserne (eds.), *National Legal Systems and Globalization* (Asser Press, 2013) 9, 23–4.

13 Carbonara et al., above n 4, 85.

14 Cheng, above n 5, 465–71; see also Arunabha Ghosh, 'Governing Clean Energy Subsidies: Why Legal and Policy Clarity Is Needed' (2011) 5(3) *Biores* <www.ictsd.org/bridges-news/biores/news/governing-clean-energy-subsidies-why-legal-and-policy-clarity-is-needed>.

were identified in the primary framework pieces of renewable energy legislation of the 113 countries with renewable energy laws. This is evidence of the heterogeneity of national preferences.

However, legal harmonisation or convergence may also impose significant costs, while legal divergence may offer some advantages. For example, negotiating and drafting harmonised legislation, domestically implementing and administering it and ensuring supranational monitoring and compliance all require significant resources.[15] This process may not only affect the primary law, but may also require changes to be made to secondary laws and regulation. If these costs are significant enough, they may even outweigh the potential benefits to be achieved by harmonising or converging, and make such a process unnecessary and/or undesirable.[16]

These costs are not just economic; they may also have strong social and political elements to them, which in turn may affect the legitimacy of the harmonised laws. Indeed, it is often argued that different national laws can better reflect the local legal traditions, customs and norms, as well as social, cultural and economic preferences.[17] They are more likely to provide a solution to the particular domestic problem at hand and to do so in a manner that maintains the legitimacy of the legislative response among their citizens.[18] In addition, Hefeker has argued that divergent national laws may also lead to greater levels of political accountability and public engagement:

> When political decisions are made at a level that is close to the population, citizens can better influence decisions and more easily hold politicians accountable who deviate too much from the electorate's interests. In addition, when people have more direct influence in policy, they are more likely to develop an interest in those policies.[19]

In order to achieve harmonised laws, political concessions are often required to win public acceptance. This may cause a number of problems that could have an impact on the effectiveness and efficiency of these laws. For example, where the resulting harmonised laws provide uniform levels of support to all countries, geographic regions, renewable energy sources and technologies,

15 House of Representatives Standing Committee on Legal and Constitutional Affairs, above n 5, 9–10; Carbonara et al., above n 4, 398–9.

16 House of Representatives Standing Committee on Legal and Constitutional Affairs, above n 5, 11–12.

17 See e.g. Carbonara and Parisi, above n 4, 370; Hefeker, above n 8, 85.

18 See e.g. Carbonara and Parisi, above n 4, 369; Bertrand Crettez, Bruno Deffains and Oliver Musy, 'On the Dynamics of Legal Convergence' (2011) 156 *Public Choice* 347.

19 Hefeker, above n 8, 85.

this will not sufficiently reflect countries' domestic contexts, leading to 'higher rents for those producers which make use of least-cost technologies and sites'.[20] Alternatively, if the harmonised support scheme permits and provides differing levels of support (and hence benefits) by country but the costs are borne equally or out of proportion to the benefits, this may lead to local opposition and a reduction in public acceptance.[21]

A further problem is that when countries harmonise their laws, there are inevitably 'winners' and 'losers'. Winning countries are those whose laws are most proximate to the regulatory standard to which the other countries are transplanting, harmonising or converging towards. This means that they gain the benefits of other countries making their laws more similar to their own with minimal costs.[22] Losing countries are those that bear the brunt of the cost of switching and adaptation. As a result, Garoupa and Ogus have argued that each country prefers its own legal rules and practices to prevail as the regulatory standard that others are harmonising to or converging towards.[23]

When countries harmonise or converge their laws to a regulatory standard, it is likely that one of two scenarios will eventuate. In the first scenario ('the race to the bottom'[24] scenario), the regulatory standard to be adopted will broadly represent the lowest common denominator and will be achieved by virtue of political compromise. Depending on the subject matter of the legislation, this may be an undesirable outcome, especially where environmental or labour standards are involved. In the context of regulatory support for renewable energy, this is likely to mean providing insufficient financial support for all but the most commercialised renewable energy sources and no certainty about the lifespan of the funding. This means that the country may be able to deploy renewable energy quickly (likely in a boom/bust cycle), but may not achieve a diverse energy mix. In the second scenario ('the race to the top' scenario), the regulatory standard to be adopted represents the 'high-water mark' of a single country. In the context of renewable energy laws, this may mean long-term favourable financial incentives that are stable and available to different energy sources and technologies at a level that reflects their relative

20 Resch et al., above n 8, 158.

21 Ibid.

22 Drahos et al., above n 3, 107–8. See also David Jacobs, *Renewable Energy Policy Convergence in the EU: The Evolution of Feed-in Tariffs in Germany, Spain and France* (Ashgate Publishing, 2012) 14.

23 Nuno Garoupa and Anthony Ogus, 'A Strategic Interpretation of Legal Transplants' (2006) 35 *The Journal of Legal Studies* 339, 347.

24 See e.g. Mads Andenæs, Camilla Baasch Andersen and Ross Ashcroft, 'Towards a Theory of Harmonisation' in Mads Andenæs and Camilla Baasch Andersen (eds.), *Theory and Practice of Harmonisation* (Edward Elgar, 2011) 572, 582.

commercialisation through the use of banding. In the short term at least, the second scenario is likely to impose significantly higher costs on less developed countries, which are less likely to have laws near the 'high-water mark'.[25]

A further advantage of legislative divergence is that it encourages regulatory innovation and encourages competition among jurisdictions to produce better and more efficient laws.[26]

6.2 THE FUTURE DEVELOPMENT OF REGULATORY SUPPORT MECHANISMS: SAME, SAME OR DIFFERENT?

As shown above, at least from a theoretical perspective, the regulatory support mechanisms used in the national renewable energy laws should become more similar over time, likely through the process of convergence. Unfortunately, other than in Europe, where there have been a large number of studies,[27] the degree of unification, harmonisation, convergence, divergence or regulatory competition within the regulatory support mechanisms in national renewable energy laws is not currently known. There are several reasons for this. First, even now, there is no comprehensive and reliable database of the national renewable energy laws for every country in the world with such laws, let alone a historical one. Second, any detailed study of all of the specific regulatory support mechanisms contained within the national renewable energy laws of

25 House of Representatives Standing Committee on Legal and Constitutional Affairs, above n 5, 10.

26 House of Representatives Standing Committee on Legal and Constitutional Affairs, above n 5, 10. See also Crettez et al., above n 18, 2.

27 Lena Kitzing, Catherine Mitchell and Poul Erik Morthorst, 'Renewable Energy Policies in Europe: Converging or Diverging?' (2012) 51 *Energy Policy* 192; Resch et al., above n 8; Jacobs, above n 22; Pablo del Río et al., 'Key Policy Approaches for a Harmonisation of RES(-E) support in Europe – Main Options and Design Elements' (Report, European IEE Project Beyond2020, March 2012); Sian Crampsie, 'Renewables Convergence?' (2011) 34(14) *Utility Week* 9; Tatiana Romanova, 'Legal Approximation in Energy: A New Approach for the European Union and Russia' in Caroline Zuzemko, Andrei V Belyi, Andreas Goldthau and Michael F Keating (eds.), *Dynamics of Energy Governance in Europe and Russia* (Palgrave Macmillan, 2012); Miquel Muñoz, Volker Oschmann and J David Tàbara, 'Harmonization of Renewable Electricity Feed-in Laws in the European Union' (2007) 35 *Energy Policy* 3104; Roger Hildingsson, Johannes Stripple and Andrew Jordan, 'Governing Renewable Energy in the EU: Confronting a Governance Dilemma' (2012) 11 *European Political Science* 18; Malgorzata Alicja Czeberkus, *Renewable Energy Sources: EU Policy and Law in Light of Integration* (LLM Thesis, University of Iceland, 2013); Per-Olof Busch and Helge Jörgens, 'Europeanization Through Diffusion? Renewable Energy Policies and Alternative Sources for European Convergence' in Francesc Morata and Israel Solorio Sandoval (eds.), *European Energy Policy* (Edward Elgar, 2012).

the 113 countries with such laws would require a large team of researchers and a considerable budget for translation. To date, this task has proven too ambitious for any research team to undertake.

Due to these limitations, the next section will examine the processes of unification, harmonisation, convergence, divergence and regulatory competition within the renewable energy sector using representative case studies to explain the likely operation of each process. At this juncture, it should also be noted that there is no agreement about the legal meaning of these concepts, with many of the definitions in the academic literature showing significant variance.[28] The discussion below will explain the definition of each concept adopted in this research as part of the analysis.

6.3 UNIFICATION

Unification, or 'hard', 'formal' or 'total harmonisation', involves all participating countries transplanting a uniform set of rules selected at an interstate level or by a supranational organisation.[29] Under unification, the relevant laws of the participating countries are effectively homogenised, with no differentiation, flexibility or derogation permitted.[30]

To date, there has not been any unification of either regulatory support mechanisms or national renewable energy laws more generally. This is not surprising, as it is exceptionally rare for countries to agree to completely unify their laws, given that it effectively amounts to them ceding their sovereignty over the issue to a supranational organisation for as long as they are a party to the unified laws. Given the lack of differentiation, flexibility or derogation, unified laws are also considered to be a hard sell politically, especially for the countries that stand to lose their comparative advantage through not having efficient and cost-effective laws tailored to their own specific contexts. This is particularly relevant if one country has to bear more of the costs for less of the benefits than another country.

28 See e.g. Eva J Lohse, 'The Meaning of Harmonisation in the Context of European Union Law – a Process in Need of Definition' in Mads Tønnesson Andenæs and Camilla Baasch (eds.), *Theory and Practice of Harmonisation* (Edward Elgar, 2011) 282; Fernando Gomez and Juan Jose Ganuza, 'How to Build European Private Law: An Economic Analysis of the Lawmaking and Harmonization Dimensions in European Private Law' (2012) 33 *European Journal of Law and Economics* 481, 483; Resch et al., above n 8, 150–1.

29 Bertrand Crettez, Bruno Deffains and Régis Deloche, 'On the Optimal Complexity of Law and Legal Rules Harmonization' (2009) 27 *European Journal of Law and Economics* 129, 131.

30 Garoupa and Ogus, above n 23, 343.

6.4 HARMONISATION

Harmonisation is a 'top-down' coercive process in which countries use over-arching legislative or other formal instrument-based mechanisms to achieve parity between legal systems.[31] This process often involves binding multilateral agreements with negotiations facilitated by a supranational organisation. These agreements specify the objectives and targets to be achieved, with countries then responsible for modifying their own internal laws to ensure those defined objectives and targets are achieved.[32]

Within Europe, countries traditionally exercised national sovereignty over the area of setting energy law and policy. Indeed, prior to the ratification of the Lisbon Treaty in 2009, the EU did not have an explicit 'shared competence' between the Union and the Member States in the field of energy.[33] Instead, it had to rely on more general shared competences, such as the internal market and environment, in order to exert indirect influence over the renewable energy sector. This lack of a specific shared competence affected the options available to the European Parliament and European Commission when they began to support the accelerated deployment of renewable energy.[34] In particular, it meant that EU Member States were able to design and implement their own regulatory support mechanisms to accelerate the deployment of renewable energy within their own countries. Even following the introduction of a 'shared competence' and a specific article on energy contained in Article 194 of the Treaty on the Functioning of the European Union (TFEU) in 2009, the EU still does not have exclusive control, due to a reservation inserted in Article 194(2) of the TFEU. This reservation enables a Member State to 'determine the conditions for exploiting its energy resources, its choice between different energy sources and the general structure of its energy supply'.[35] It is within this context that there have thus far been two

[31] House of Representatives Standing Committee on Legal and Constitutional Affairs, above n 5, 1.

[32] Carbonara et al., above n 4, 368.

[33] Now *Treaty on European Union*, opened for signature 7 February 1992, [2009] OJ C 115/13 (entered into force 1 November 1993) Art 4; *Treaty on the Functioning of the European Union*, opened for signature 7 February 1992, [1992] OJ C 115/199 (entered into force 1 November 1993) Art 4(2)(i).

[34] Jacobs, above n 22, 29.

[35] *Treaty on the Functioning of the European Union*, opened for signature 7 February 1992, [1992] OJ C 115/199 (entered into force 1 November 1993) Art 194(2). It is not yet known how this Article may be interpreted: see e.g. Angus Johnston and Eva van der Marel, '*Ad Lucem?* Interpreting the New EU Energy Provision, and in Particular the Meaning of Article 194(2) TFEU' (2013) *European Energy and Environmental Law Review* 181; Kristín Haraldsdóttir, 'The Limits of EU Competence to Regulate Conditions for Exploitation of Energy

unsuccessful attempts to harmonise the regulatory support mechanisms used in the renewable energy sector within the EU. Each of these attempts will be discussed below.

6.4.1 *The First EU Harmonisation Attempt*

The origins of the first attempt to harmonise the support mechanisms within the EU are found within the 1995 Energy White Paper[36] and the 1996 Green Paper, 'Energy for the Future'. The latter document argued for 'a stable and Community wide framework for renewable energy sources'.[37] Then, in 1997, the Commission outlined the first common European policy strategy dealing with renewable energy, in which it proposed the establishment of a non-legally binding goal of doubling the share of renewable energy to 12 per cent by 2010.[38] During this period, there was intense discussion within the EU about whether harmonisation was appropriate and, if so, what form it should take. Proponents of harmonisation argued that it was necessary to avoid market distortions caused by having national regulatory support schemes, that it would support the development of a European internal energy market and be more efficient and cost effective.[39] Meanwhile, opponents argued that harmonisation would lead to higher costs and could also prove disruptive to the development of the European renewable energy sector.[40]

The debate came to a head following the release of the first draft of what would later become the Renewable Energy Directive in October 1998.[41] The draft Directive proposed that a European-wide quota-based green certificate scheme be established and that FITs be prohibited.[42] This proposal was strongly supported by some of the early adopters of green certificate schemes,

Resources: Analysis of Article 194(2) TFEU' (2014) *European Energy and Environmental Law Review* 208.

36 Commission of the European Communities, *An Energy Policy for the European Union*, COM (95) 682 Final (13 December 1995).

37 Commission of the European Communities, *Energy for the Future: Renewable Sources of Energy*, COM(96) 576 Final (20 November 1996) 28.

38 European Commission, *Energy for the Future: Renewable Sources of Energy – White Paper for a Community Strategy and Action Plan*, COM(97) 599 Final (26 November 1997) 10.

39 Jacobs, above n 22, 36–7; Secretary-General of the European Commission, *Electricity from Renewable Energy Sources and the Internal Electricity Market*, SEC(1999) 470 Final (13 April 1999).

40 Jacobs, above n 22, 38.

41 Francesc Morata and Israel Solorio Sandoval (eds.), *European Energy Policy* (Edward Elgar, 2012) 75–6.

42 Ibid.

such as the United Kingdom.[43] However, it also faced staunch opposition from both Germany and Spain, which both had FIT schemes in place.[44] This opposition was compounded in 2001 when the European Court of Justice held in *PreussenElektra* v. *Schleswag*[45] that the German feed-in law, which the Commission had referred to the Court for its assumed breach of European competition law and the principles governing the liberalisation of the European electricity market, was not incompatible with EU law.

These factors led to the final version of the 2001 Directive on Electricity Production from Renewable Energy Sources[46] postponing harmonisation of the support schemes until 2012. Instead, it fixed indicative renewable energy targets for each Member State to achieve by 2010 but, in accordance with the principle of subsidiarity, Member States were able to select the most appropriate support mechanism to achieve those targets. The support schemes adopted by Member States were to be reviewed in accordance with Article 4 of the Directive by 2005, and, if necessary, be 'accompanied by a proposal for a Community framework with regard to support schemes for electricity produced from renewable energy sources'.[47] It was this review that played a key role in the failure of the second opportunity for the EU to harmonise the support schemes across all Member States.

While this first attempt at harmonisation ultimately failed, research conducted by Morata and Sandoval has indicated that the support for green certificate schemes by the Commission and the political uncertainty around the future of FITs led to at least two countries (Denmark and Holland) adopting a green certificate scheme.[48]

6.4.2 The Second EU Harmonisation Attempt

In December 2005, the review of the support schemes adopted by the Member States to support renewable electricity was delivered.[49] This review had two key conclusions:

43 Hildingsson et al., above n 27, 21.
44 Ibid; Morata et al., above n 41, 79.
45 [2001] EUECJ C-379/98.
46 *Directive 2001/77/EC of the European Parliament and of the Council of 27 September 2001 on the promotion of electricity produced from renewable energy sources in the internal electricity market* [2001] OJ L 283.
47 Ibid Art 4(2).
48 Morata et al., above n 41, 78.
49 European Commission, *The Support of Electricity from Renewable Energy Sources*, COM (2005) 627 Final (7 December 2005).

1. that FITs were more efficient than quota-based green certificate systems in deploying new renewable generation capacity; and
2. that FITs were also more cost-effective than quota-based green certificate systems.

This second conclusion was unexpected, and effectively quashed any plans in the short to medium term to harmonise a quota-based green certificate scheme across Europe.[50] Further, in the review, the Commission described the potential harmonisation of feed-in tariffs as 'difficult'[51] due to the problems associated with effectively pricing a Europe-wide FIT. It therefore chose not pursue that course of action. Instead, the Commission advocated that countries should optimise their national systems and intensify cooperation between Member States.[52]

In 2007, a target of 20 per cent of energy consumption to come from renewable energy sources by 2020 was approved.[53] This became a central tenet of the 2009 Directive, which abrogated the 2001 Directive.[54] In an early leaked draft of the 2009 Directive, the Commission proposed unrestricted certificate trading in guarantees of origin (GO).[55] This was strongly opposed by a number of countries, particularly those with FIT schemes, which they felt would be undermined by this move.[56] This opposition led the second attempt to at least partly harmonise the support schemes to fail. This point is made explicitly by Recital 25 of the Directive, which states:

> Member States have different renewable energy potentials and operate different schemes of support for energy from renewable sources at the national level . . . For the proper functioning of national support schemes it is vital that Member States can control the effect and costs of their national support

50 Muñoz et al., above n 27, 3105.

51 European Commission, *The Support of Electricity from Renewable Energy Sources*, COM (2005) 627 Final (7 December 2005) 4.

52 Ibid 16.

53 European Renewable Energy Council, *Renewable Energy in Europe: Markets, Trends and Technologies* (Routledge, 2010) 4.

54 Directive 2009/28/EC of the European Parliament and of the Council of 23 April 2009 on the promotion of the use of energy from renewable sources and amending and subsequently repealing Directives 2001/77/EC and 2003/30/EC (Text with EEA relevance) [2009] OJ L 140.

55 Jacobs, above n 22, 34. A 'guarantee of origin' means an electronic document which has the sole function of providing proof to a final customer that a given share or quantity of energy was produced from renewable sources as required by Article 3(6) of Directive 2003/54/EC: Directive 2009/28/EC of the European Parliament and of the Council of 23 April 2009 on the promotion of the use of energy from renewable sources and amending and subsequently repealing Directives 2001/77/EC and 2003/30/EC (Text with EEA relevance) [2009] OJ L 140, Art 2(j).

56 Ibid.

> schemes according to their different potentials. One important means to achieve the aim of this Directive is to guarantee the proper functioning of national support schemes . . .[57]

Instead of unrestricted GO trading, in the final 2009 Directive, Member States were able to engage in 'flexible cooperation mechanisms' such as the statistical transfer of renewable energy produced in excess of their 'mandatory' national target[58] to other Member States.[59] They were also now permitted to participate in joint projects and support schemes.[60] These cooperative mechanisms are discussed further at 6.5.3.1.

6.4.3 *A Third EU Harmonisation Attempt by Stealth?*

Despite the two previous attempts to harmonise having failed, and a general shift towards cooperation and coordination, a number of experts have now stated that the European Commission has attempted yet again to harmonise the regulatory support mechanisms by stealth using the State Aid Guidelines.[61] Prior to the current Guidelines being drafted, in 2013 the Commission had reviewed the implementation of the 2009 Directive criticising the ensuing development of

> . . . rigid national support schemes (that) were generally unable to adapt rapidly enough to such falling costs, raising profits and creating a rate and scale of installations in some countries almost excessive in a time of general economic crisis.[62]

57 Directive 2009/28/EC of the European Parliament and of the Council of 23 April 2009 on the promotion of the use of energy from renewable sources and amending and subsequently repealing Directives 2001/77/EC and 2003/30/EC *(Text with EEA relevance)* [2009] OJ L 140.

58 The national targets vary by Member State, taking into account their existing energy mix and resources. The combined targets equate to a 20 per cent EU-wide target for renewable energy production in 2020.

59 Directive 2009/28/EC of the European Parliament and of the Council of 23 April 2009 on the promotion of the use of energy from renewable sources and amending and subsequently repealing Directives 2001/77/EC and 2003/30/EC (Text with EEA relevance) [2009] OJ L 140 Art 6.

60 Directive 2009/28/EC of the European Parliament and of the Council of 23 April 2009 on the promotion of the use of energy from renewable sources and amending and subsequently repealing Directives 2001/77/EC and 2003/30/EC *(Text with EEA relevance)* [2009] OJ L 140 Arts 7–11.

61 See e.g. Kerstin Tews, 'Europeanization of Energy and Climate Policy: The Struggle Between Competing Ideas of Coordinating Energy Transitions' (2015) 24(3) *Journal of Environment & Development* 267; Tim Maxian Rusche, *EU Renewable Electricity Law and Policy: From National Targets to a Common Market* (Cambridge University Press, 2015) 224–5.

62 European Commission, *Commission Staff Working Document: Accompanying the Document: Report from the Commission to the European Parliament and the Council: Renewable energy progress report* (COM(2013) 175 final) (SWD(2013) 102 final) 27 March 2013, 5.

The Commission stated that they would formulate State Aid guidelines to help Member States identify best practice in renewable energy support schemes 'to ensure convergence and the Europeanisation of energy'.[63]

State Aid is defined in Art 107(1) of the TFEU as 'any aid granted by a Member State or through State resources in any form whatsoever which distorts or threatens to distort competition by favouring certain undertakings or the production of certain goods. The default position under Article 107 of the TFEU is that all State Aid is prohibited in so far as it affects trade between Member States on the grounds that it is incompatible with the internal market unless it fulfils the requirements of Art 107(3). In addition to the existing State Aid law, on 28 June 2014, the European Commission adopted the 'Guidelines on state aid for environmental protection and energy 2014–2020'. The adoption of these Guidelines prompted some critics to state that the Commission was 'using competition law to shape energy policy',[64] and query whether the Guidelines violated the Member States' shared competence for energy law under Art 194 of the TFEU.[65] This is because unlike in the area of energy where the Commission shares competence with Member States, pursuant to Art 108, the Commission has exclusive competence to assess whether State Aid measures are compatible with the internal market.

The State Aid Guidelines stipulate that the Member States should adopt market-based systems, in particular, from 2017 onwards; they should use competitive bidding and market premiums as their default mechanism (though some very limited grounds for an opt-out option still exist). This has had a noticeable impact on the design of new regulatory support schemes, with far greater use of competitive bidding, feed-in premiums and opening up national support schemes to foreign producers among Member States.[66] However, rather than being an example of harmonisation, the actual enforcement of the Guidelines has been more muted, due to their non-binding nature. Talus and Marhold have stated that this is in line with EU law making, preferring soft harmonisation and convergence through the use of voluntary systems of rules (and the sharing of best practice through them) before creating mandatory and legally binding EU regulation.[67]

63 Ibid 9.

64 Tews, above n 61, 267–91, 277.

65 Ibid. See also Daniel Behn, Ole Kristian Fauchald and Laura Létourneau-Tremblay, 'Promoting Renewable Energy in the EU: Shifting Trends in Member State Policy Space' (2017) 28(2) *European Business Law Review* 217.

66 See e.g. European Commission, 'State Aid: Commission Approves Hungarian Support Scheme for Renewable Electricity' (Press Release, 11 July 2017).

67 Anna-Alexandra Marhold, 'EU State Aid Law, WTO Subsidy Disciplines and Renewable Energy Support Schemes: Disconnected Paradigms in Decarbonizing the Grid' TILEC,

Thus, to date, all attempts to internationally harmonise regulatory support mechanisms within the EU have failed. Despite this, research suggests that there has been a degree of convergence in the support schemes established in EU Member States. It appears that prior to 2014, this convergence occurred in spite of, rather than because of, the European Commission's attempts to harmonise the schemes. In particular, research suggests that the convergence was around the uptake of FITs rather than green certificates, which up until recently had been the European Commission's preferred form of support scheme. This, coupled with the addition of the reservations contained in Article 194(2) of the TFEU, suggests that countries guard their ability to be able to make their own decisions with regard to energy supply. As a result, for as long as countries seek different objectives in their national renewable energy laws, they are unlikely to agree to harmonised regulatory support mechanisms. Instead, it is likely that the EU Member States will allow their laws to slowly converge, while retaining the ability to tailor their national schemes to meet their domestic needs.

6.5 CONVERGENCE

Convergence, or 'soft', 'informal' or 'minimum harmonisation', is a 'bottom-up' voluntary process in which the laws and regulations of different countries become more similar over time in a given field. It is premised on the belief that the common set of underlying principles adopted is 'flexible enough to be adapted to countries under disparate socio-economic circumstances'.[68] The primary goal of convergence is to facilitate similar levels of market access to investors and other participants within the countries in order to encourage trade and commerce.[69]

There are four key ways in which laws may converge. First, convergence may occur unilaterally, with a country independently transplanting the 'statutes and principles belonging to other systems, be they legal rules of other countries or customs whose acceptance is widespread'.[70] An example of this

Discussion Paper 2017–029 (Tilberg University, July 2017) 17; Kim Talus, *Introduction to EU Energy Law* (Oxford University Press, 2016) 124.

68 Cheng, above n 5, 445.

69 Anatole Boute, 'Improving the Climate for European Investments in the Russian Electricity Production Sector: (II) the Role of Regulatory Convergence' (2008) 26 *Journal of Energy & Natural Resources Law* 327, 333.

70 Carbonara et al., above n 4, 368. See also Karoline Steinbacher, 'Drawing Lessons When Objectives Differ? Assessing Renewable Energy Policy Transfer from Germany to Morocco' (2015) 3(2) *Politics and Governance* 34.

may be found in the Malawian Energy Regulation Act 2004. The legislative drafters of this Act have clearly transplanted the definition of 'renewable energy' from another country due to their inclusion of wave energy in their definition despite Malawi being landlocked and not part of a regional electricity market. As such, this is an example of the convergence of definitions of 'renewable energy' on paper but where it will not have any practical effect. Second, two countries may simultaneously devise and implement the same legal solution to the same problem with no knowledge of the other country's efforts.[71] Third, one country may emulate the solutions or laws of another country, learning from the experience of more experienced jurisdictions.[72] This process is usually assisted by communication between the governments, legislators, lawyers, judiciary, public servants and/or legislative drafters of the two countries.[73] Holzinger et al. have stated that 'countries that share a common language, common borders, and common traditions are more likely to adopt similar policies'.[74] However, Jordan et al. have stated that 'the policy convergence literature suggests that states following similar pathways of social and economic development will naturally gravitate towards common policies and policy instruments', though they do also note that 'societal convergence is highly contested'.[75] This process may also be facilitated by regional or international organisations through the issuance of best practices and recommendations.[76] Finally, participating countries may use cooperative and coordinated adaptation processes to encourage diverse legal rules and traditions to converge. In this context, cooperation refers to joint efforts between countries to help achieve their national renewable energy objectives.[77] Coordination, meanwhile, refers to the exchange of information between governments that may lead over time to improved knowledge about the design and implementation of regulatory support mechanisms. The use of the term 'coordination' within the context of the EU has its own special meaning and refers to the 'Open Method of Coordination'.[78]

71 Jacobs, above n 22, 14.

72 Cheng, above n 5, 441, 480. See also Steinbacher, above n 70; Gomez et al., above n 28, 484–5; Jordan et al., above n 4, 7.

73 Gomez et al., above n 28, 486.

74 Katharina Holzinger, Christoph Knill and Thomas Sommerer, 'Environmental Policy Convergence: The Impact of International Harmonization, Transnational Communication, and Regulatory Competition' (2008) 62 *International Organization* 553, 582.

75 Jordan et al., above n 4, 6.

76 Cheng, above n 5, 445.

77 Busch et al., above n 27, 68.

78 Resch et al., above n 8, 151–2.

6.5.1 *Joint Support Schemes: The Swedish-Norwegian Electricity Certificate Market*

Following ongoing opposition towards its attempts to harmonise, the EU has moved towards coordination and cooperation between Member States, focusing on the development of best practices. As stated above, one method of cooperation between Member States is the ability to establish Joint Support Schemes in accordance with Articles 10 and 11 of the 2009 Directive.[79] One example of such a scheme is the Joint Swedish-Norwegian Electricity Certificate Market, which was established on 1 January 2012 in order to increase the combined renewable electricity production in both countries by 28.4TWh by the end of 2020,[80] so that they can meet their goals under the 2009 Directive. This joint scheme stipulates that 'Sweden will finance 15.2 TWh and Norway will finance 13.3 TWh, but it is up to the market to decide where and when the new production is to take place'.[81] For example, during 2016, almost 60 per cent of the anticipated production came from Swedish wind power.[82] The Swedish-Norwegian Electricity Certificate Market appears to be an effective model of cooperation between Member States, with 15.5TWh of new renewable energy production capacity built in the period from 1 January 2012 to 30 April 2016.[83]

The European Court of Justice in its judgment in *Ålands Vindkraft AB v. Energimyndigheten*[84] confirmed that the existence of these cooperative measures within the 2009 Directive does not require Member States to permit renewable energy producers generating electricity within the confines of another Member States to participate in their national support schemes. This means that unless the Member States formally agree to participate in a Joint Support Scheme, Member States may elect to only provide their regulatory support mechanism to renewable energy projects located within their borders, providing this limitation is justified on environmental grounds.

79 Directive 2009/28/EC of the European Parliament and of the Council of 23 April 2009 on the promotion of the use of energy from renewable sources and amending and subsequently repealing Directives 2001/77/EC and 2003/30/EC (Text with EEA relevance) [2009] OJ L 140 Arts 10–11.

80 Swedish Energy Agency and the Norwegian Water Resources and Energy Directorate, 'Joint Swedish-Norwegian Electricity Certificate Market Annual Report 2016' (Report, 2016) Preface.

81 Ibid.

82 Ibid 6.

83 Reuters, 'Norway, Sweden to Meet Green Energy Target Under Joint Scheme', 9 June 2016 <www.reuters.com/article/norway-sweden-renewables-idUSL8N1911TK>.

84 (C573-12) [2014] ECR 2037.

This approach by the ECJ has been upheld in subsequent cases, including the *Essent Belgium* case.[85]

Other than through Joint Support Schemes within the EU, the full extent of convergence is unknown. For example, while it is clear that countries are now benefiting from coordinated knowledge sharing through bodies such as the IRENA, the IEA and regional organisations, or by emulating best practices in regulatory support mechanisms, the impact is difficult to measure. There is also evidence that countries are seeking to learn from the regulatory and market failures of other countries.[86] Given the recent spate of investor–state disputes that are currently being arbitrated under the Energy Charter Treaty, and/or other bilateral or multilateral investment agreements, this is also likely to be a fertile source of knowledge that countries may seek to converge around.[87] It is hypothesised that, in particular, the fifty-two Member and thirty-four Observer Countries of the ECT will use the arbitral decisions from cases such as *Charanne*,[88] *Isolux*[89] and *Eiser*[90] to learn how to legally modify their regulatory support mechanisms and not fall foul of the legitimate expectations and fair and equitable treatment provisions of the Treaty. However, as the vast majority of these cases are still pending, the extent of this convergence is difficult to quantify and requires further research.

6.6 DIVERGENCE

The laws of a country usually incorporate the local preferences of its citizenry, as well as its unique political, economic, social, legal and environmental contexts. This process reinforces the legitimacy of a country's national laws and explains why countries adopt differing legal solutions to similar

85 Joined Cases C-204/12 to C-208/12, *Essent Belgium NV* v. *Vlaamse Reguleringsinstantie voor de Elektriciteitsen Gasmarkt* [NYR] (European Court of Justice, 11 September 2014).

86 Steinbacher, above n 70.

87 See e.g. Yulia S. Selinova, 'Changes in Renewables Support Policy and Investment Protection under the Energy Charter Treaty: Analysis of Jurisprudence and Outlook for the Current Arbitration Cases' (2018) *ICSID Review* 1; Daniel Behn et al., above n 65, 233–8; Martin Švec, 'The Energy Charter Treaty: Renewable Energy Disputes in Light of the Charanne Case' in *Cofola International 2016: Resolution of International Disputes and Public Law in the Context of Immigration Crisis: Conference Proceedings* (Masaryk University, 2016) 237.

88 *Charanne B.V. and Construction Investments S.a.r.l.* v. *Spain (Award)* (SCC Arbitral Tribunal, SCC Case No. 062/2012, 21 January 2016).

89 *Isolux Infrastructure Netherlands B.V.* v. *Kingdom of Spain (Award)* (SCC Arbitral Tribunal, SCC Case No. 2013/153, 17 July 2016).

90 *Eiser Infrastructure Limited and Energía Solar Luxembourg S.à.r.l.* v. *Kingdom of Spain (Award)* (ICSID Arbitral Tribunal, ICSID Case No. ARB/13/36, 4 May 2017).

problems.[91] In fact, it could be argued that, in the absence of economic or political interdependencies between countries or the presence of international externalities, the divergence of laws should be the natural state of affairs. While there are costs involved in countries adopting different national renewable energy laws, largely in the form of transaction costs for international market participants, there are also benefits to divergence. The benefits of divergence are most obvious when the differing national laws reflect local preferences.

There are a number of ways in which laws addressing the same problem may diverge: conceptual divergence (differing definitions of fundamental terms); normative divergence (differing underlying principles, goals or objectives); substantive divergence (differing legal instruments and mechanisms); and procedural divergence (differing administrative procedures regarding implementation and interpretation). All of these forms of divergence have the potential to have an impact on the design and operation of regulatory support mechanisms.

It is difficult to know the extent of this phenomenon without a comprehensive longitudinal study of the regulatory support mechanisms contained within the national renewable energy laws of every country that has such laws. However, in light of the above, it is likely that the starting position among the majority of countries with national renewable energy laws appears to be one of divergence (the exception being the Member States and candidate countries of the EU).

6.7 REGULATORY COMPETITION

Theories of regulatory competition argue that, in the context of industries subject to economic market integration and free trade,[92] national legislation becomes a competitive parameter.[93] In this environment, governments face pressure to reduce or remove the regulatory burdens that may impair the competitiveness of economic actors operating within their jurisdiction to avoid these actors moving elsewhere.[94] Thus, regulatory competition has

91 Chirico et al., above n 12, 14.

92 Jacobs, above n 22, 219.

93 Hanne Søndergaard Birkmose, 'Regulatory Competition and the European Harmonisation Process' (2006) 17 *European Business Law Review* 1075, 1076; see also Dale D Murphy, 'The Puzzle and an Explanation' in Dale D Murphy (ed.), *The Structure of Regulatory Competition: Corporations and Public Policies in a Global Economy* (Oxford University Press, 2006) 4.

94 Holzinger et al., above n 74, 560.

been defined as 'the process whereby regulators deliberately set out to provide a more favourable regulatory environment in order to improve the competitiveness of domestic industries or to attract more business activity from abroad'.[95]

In the short to medium term, regulatory convergence may be said to lead to active divergence as countries actively seek to compete via their legislation. There are three possible outcomes for this process:

1. Countries may become more protectionist in order to support their domestic industries by increasing the regulatory barriers for foreign competitors to enter the national market.[96] 'This is often coupled with industry support in the form of regulatory support mechanisms for domestic industry participants. This is likely to lead to higher domestic prices and may also pose difficulties for exporters if other countries adopt a tit-for-tat approach.'[97] This approach to regulatory competition is also likely to result in international trade conflicts for potential breaches of the General Agreement on Tariffs and Trade (GATT), SCM Agreement and anti-dumping laws.
2. Countries may seek to lower the level or standards of their regulation in order to make themselves a more attractive destination for foreign investors. If this competition-in-laxity is sustained, it may lead to a 'race to the bottom' with countries competing to introduce lower and weaker standards.[98]
3. Countries may seek to increase the regulatory support mechanisms available to producers in a 'race to the top'.[99] These mechanisms may target a number of areas in an attempt to 'increase the probability that a locally preferred design becomes internationally successful',[100] such as innovation, regulation, market structures, competitiveness and export orientation.[101] Often countries adopting this approach will be seeking to become the lead market for the product.[102] This approach is most visible in international markets where demand outweighs supply, and/or

95 Birkmose, above n 93, 1076.

96 See e.g. Holzinger et al. above n 74, 560; Murphy, above n 93, 12.

97 Murphy, above n 93, 13.

98 See e.g. Dale D Murphy, 'Evidence and Implications' in Dale D Murphy (ed.), *The Structure of Regulatory Competition: Corporations and Public Policies in a Global Economy* (Oxford University Press, 2006) 216; Birkmose, above n 93, 1078.

99 Mario Ragwitz et al., 'EmployRES: The Impact of Renewable Energy Policy on Economic Growth and Employment in the European Union' (Report, Karlsruhe, 27 April 2009) 10.

100 Ibid.

101 Ibid.

102 Ibid.

a country has a strategic reason such as economic or security concerns for seeking to accelerate their deployment. In the long run, this becomes uneconomic because due to financial restrictions countries cannot continue to increase the levels of support offered.

In the longer term, regulatory competition actually leads to the convergence of laws because in an international market the countries that are competing with one another are likely to be subject to similar market pressures and respond in similar ways.[103] For example: country A has a national renewable energy law that is perceived to provide a strong competitive market for investment, country B wants to enact a national renewable energy law, so it copies the legislation of country A, but makes minor improvements. When country A goes to review and amend its laws, it may look to the improvements made by country B and adopt them with its own minor improvements. Over time, this will lead to their laws becoming more similar, leading Gomez et al. to state that it leads to 'a sort of competitively harmonized legal regime'.[104] A number of empirical studies suggest that the *direction* of the convergence – that is, whether it is going to be a 'race to the bottom' or a 'race to the top' – is less predictable.[105]

6.7.1 *International Trade Conflicts Arising from National Renewable Energy Laws*

Within the renewable energy sector, regulatory competition is almost always coupled with interventions justified on the grounds of industrial policy. Industrial policy has been described as 'a strategy to revitalize, improve, and develop an industry',[106] and 'a set of policies that selectively favours the development of certain industries over others'.[107] The rationale for the use of industrial policy is that it enables governments to correct market failures, which are encumbering the development of new industries, and the research and development, commercialisation and widespread adoption of new technologies.[108]

103 See e.g. Holzinger, above n 74, 561; Anthony Ogus, 'Competition Between National Legal Systems: A Contribution of Economic Analysis to Comparative Law' (1999) 48 *International and Comparative Law Quarterly* 405, 407–8.

104 Gomez et al., above n 28, 485–6.

105 Jacobs, above n 22, 15.

106 Bob Carbaugh and Max St Brown, 'Industrial Policy and Renewable Energy: Trade Conflicts' (2012) 5(1) *Journal of International and Global Economic Studies* 1, 2.

107 Johannes Schwarzer, 'Industrial Policy for a Green Economy' (Report, International Institute for Sustainable Development, June 2013) iii.

108 Carbaugh et al., above n 106, 1–2.

Within the renewable energy sector, industrial policy commonly takes the form of direct market interventions within the national economy to foster the development of indigenous renewable energy sources, support domestically designed and manufactured technologies and to encourage job creation.[109] These market interventions can take a number of forms, including local content clauses, import tariffs, the provision of tax incentives, low interest rate loans, loan guarantees and subsidies. These interventions act to reduce prices for consumers, increase the prices paid to renewable energy technology manufacturers or generators of electricity from renewable energy sources or reduce the cost of production.[110] Almost invariably, these interventions favour domestic producers at the expense of foreign competitors, based on the belief that by constraining:

> the spread of technology knowledge to foreign countries ... the restrictions on transfer of innovation will benefit national economic growth, protect national wealth, and secure energy independency.[111]

This conduct is particularly noticeable in the renewable energy sector when it was in its market infancy and development phases. Between 2005 and 2013, no fewer than fourteen countries, including China, Brazil, India, the United States, France, Italy and Spain, inserted local content requirements into their regulatory support mechanisms in an attempt to foster the development of their domestic industries.[112]

The development of national renewable energy laws that favour domestic production over foreign imports on the basis of industrial policy actively hinders the operation of the free market economy. This prompts two common criticisms. First, doubts have been cast over the ability of the government to correct market failures and to deliver a comparative advantage to their domestic renewable energy technologies by picking winners.[113] Second, due to the protectionist nature of these laws, they almost invariably

[109] See e.g. Lewis, above n 1, 11.

[110] Carbaugh et al., above n 106, 1–2.

[111] Catherine Banet, 'Techno-nationalism in the Context of the Energy Transition' in Donald Zillman, Lee Godden, LeRoy Paddock and Martha Roggenkamp (eds.), *Innovation in Energy Law and Technology: Dynamic Solutions for Energy Transitions* (Oxford University Press, 2018) 75.

[112] See e.g. Vyoma Jha, 'Political Economy of Climate, Trade and Solar Energy in India' (2017) IX(2) *Trade, Law and Development* 155.

[113] See e.g. Carbaugh et al., above n 106, 4, 12; Paolo D Farah and Elena Cima, 'Energy Trade and the WTO: Implications for Renewable Energy and the OPEC Cartel' (2013) 16 *Journal of International Economic Law* 707, 726.

lead to international disputes about whether they breach WTO law and free trade agreements.[114]

In recent years, there have been a number of referrals of national renewable energy laws that impose unfair trade barriers on imports to organisations such as the WTO, Chinese Ministry of Commerce, European Commission, US Department of Commerce/International Trade Commission and the Indian Ministry of Commerce.[115] The rise of international trade disputes involving the renewable energy sector is a relatively recent but rapidly growing phenomenon. These disputes cover a range of trade barriers such as local content requirements, anti-dumping, countervailing measures and the provision of subsidies. Unfortunately, due to the time taken for the WTO to hear disputes and the propensity for countries to settle them there is often limited information available about the current status of the disputes. However, one pattern that is clear is that China is the most common respondent to these claims. To highlight how these cases reflect regulatory competition and industrial policy, a case study of the Chinese wind subsidy dispute will be discussed below.

6.7.2 *Case Study: Chinese Wind Subsidy WTO Dispute*

China has often been accused of using its renewable energy regulations to promote its clean technology industry, which has been identified as a key growth industry for the twenty-first century.[116] In 2009, Premier Wen Jiabao explicitly stated:

> We will accelerate the development of a low-carbon and green economy so as to gain an advantageous position in international industrial competition.[117]

This ambition has only intensified over the years, with the Made in China 2025 Report stating that China aims to produce 80 per cent of the global supply of

114 Lewis, above n 1, 16; Carbaugh et al., above n 106, 2; Anton Ming-Zhi Gao, 'Promotion of Renewable Electricity: Free Trade and Domestic Industrial Development' in Kim Talus (ed.), *Research Handbook on International Energy Law* (Edward Elgar, 2014) 408.

115 See e.g. Marie Wilke, 'Feed-in Tariffs for Renewable Energy and WTO Subsidy Rules: An Initial Legal Review' (Issue Paper No. 4, Institutional Centre for Trade and Sustainable Development, November 2011).

116 Chien-Huei Wu and Kuei-Chih Yang, 'Aggressive Legalism: China's Proactive Role in Renewable Energy and Trade Disputes?' (2015) 12(3) *Transnational Dispute Management* 1; Daniel Yergin, *The Quest: Energy, Security and the Remaking of the Modern World* (Penguin Press, 2011) 544–5.

117 Wen Jiabao, 'Strengthen Confidence and Work Together for a New Round of World Economic Growth' (Speech delivered at the World Economic Forum, Switzerland, 28 January 2009) quoted in Yergin, above n 116, 544.

'basic core components and important basic materials' for renewable energy equipment by 2025.[118]

This conduct has led to complaints that the Chinese are using their renewable energy regulations, and, in particular, those directed at providing subsidies, to enable Chinese manufacturers to be more competitive with foreign companies.[119] In 2005, the market share of foreign turbine firms in China was 75 per cent.[120] To counter this growth, 'the National Development and Reform Commission (NDRC) introduced a cap that required Chinese wind farms to source at least 70% of turbine parts from domestic producers'.[121] This regulatory policy was so effective that, over a three-year period, the market share of foreign turbine firms had declined by 55 percentage points to a 20 per cent market share.[122] Over this period, China also went from having only six domestic wind turbine manufacturers to being the number one producer of wind turbines in the world in 2009.[123] Following international pressure, the domestic content cap was subsequently revoked,[124] only to be replaced a few years later by conditional subsidies that gave preferential treatment to domestic wind turbine manufacturers.

In 2011, the US Government on behalf of the United Steelworkers consulted the WTO over the Chinese Special Fund for Wind Power Equipment Manufacturing subsidies,[125] which they alleged breached Article 3 of the WTO's Agreement on Subsidies and Countervailing Measures ('SCM').[126] The relevant clauses of this Article state:

> 3.1 … the following subsidies, within the meaning of Article 1, shall be prohibited:

[118] Lily Kuo, 'Made in China policy at centre of tariff war with US', *The Guardian*, 4 April 2018 <www.theguardian.com/world/2018/apr/04/made-in-china-policy-at-centre-of-tariff-war-with-us>.

[119] Yergin, above n 116, 544–5.

[120] Vinod K Aggarwal and Simon Evenett, 'The Financial Crisis, "New" Industrial Policy, and the Bite of Multilateral Trade Rules' (2010) 5 *Asian Economic Policy Review* 221, 234.

[121] Ibid.

[122] Ibid.

[123] Ibid 232.

[124] Ibid 234.

[125] 风力发电设备产业化专项资金管理暂行办法 [Management Regulations on Special Fund for the Industrialization of Wind Power Manufacturing Sector in China] (Ministry of Finance Document No. 476, People's Republic of China, 11 August 2008).

[126] Office of the United States Trade Representative: Executive Office of the President, 'China Ends Wind Power Equipment Subsidies Challenged by the United States in WTO Dispute' (Press Release, 6 June 2011) <https://ustr.gov/about-us/policy-offices/press-office/press-releases/2011/june/china-ends-wind-power-equipment-subsidies-challenged>.

(b) subsidies contingent, whether solely or as one of several other conditions, upon the use of domestic over imported goods.

3.2 A Member shall neither grant nor maintain subsidies referred to in paragraph 1.[127]

The Chinese Special Fund for Wind Power Equipment Manufacturing subsidies were explicitly designed to support the domestic research and development of MW-scale wind turbine systems in China. The qualifying criteria for this subsidy contained local content clauses, including:

- to be eligible companies must be State-owned or Chinese-controlled wind power equipment manufacturers (including wind turbine and component manufacturers);
- the developed equipment must have the Chinese Intellectual Property Right (IPR), i.e. the company must own the critical technology or techniques; and
- the wind turbine systems must be manufactured, installed and tested in China and must be operated without fault for more than 240 hours.[128]

This meant that, in contravention of Articles 3.1(b) and 3.2 of the SCM, the subsidies were providing preferential financial assistance to Chinese wind turbine manufacturers that used domestic components and manufacturing rather than purchasing imports.[129] Given that individual grants ranged between $US6.7 million and $US22.5 million, it has been estimated that the total value of the subsidy given to Chinese manufacturers may have totalled several hundred million dollars between 2008 and 2011.[130] Following the WTO consultations between the United States and China on 11 February 2011, China formally revoked the Management Regulations on the Special Fund.[131]

The problems experienced by foreign companies trying to gain access to the Chinese wind market are not unique. Companies working in the renewables sector within the region have often complained about the overt discrimination displayed towards foreign firms trying to gain a foothold in some of the world's

[127] *Marrakesh Agreement Establishing the World Trade Organization*, opened for signature 15 April 1994, [1994] 1867 UNTS 3 (entered into force 1 January 1995) annex 1A Arts 3.1–2.

[128] 风力发电设备产业化专项资金管理暂行办法 [Management Regulations on Special Fund for the Industrialization of Wind Power Manufacturing Sector in China] (Ministry of Finance Document No. 476, People's Republic of China, 11 August 2008).

[129] Office of the United States Trade Representative: Executive Office of the President, above n 126.

[130] Ibid.

[131] Ibid.

largest renewable electricity markets.[132] Indeed, foreign wind turbine manufacturers such as the Indian wind turbine manufacturer, Suzlon,[133] have complained that they have been hamstrung by the frequent regulatory changes,[134] as well as a lack of transparency around the decision-making processes being applied to bidding rounds for projects.[135] It would seem that explicit divergence in the form of regulatory competition in order to further national industrial policy is prevalent within the renewable energy sector, particularly among those countries that are active as technology innovators.

6.8 CONCLUSION

In a sector that has experienced substantial growth both in investment and in installed capacity over the past fifteen years, the benefits of the regulatory support mechanisms within national renewable energy laws becoming more similar are clearly evident, especially for international market participants. Greater similarity in national renewable energy laws should lead to lower transaction costs, greater competition and, ultimately, lower prices for the ultimate consumers.

The issue of whether regulatory support mechanisms are growing more similar or more divergent over time requires further research. However, from the case studies detailed above, some patterns are evident. The starting position for most countries seems to be one of substantive divergence, with different regulatory support mechanisms being designed and implemented in different countries. This is a natural response to their different natural resources, legal traditions, governmental and socio-economic structures and customs and norms, which in turn reinforces the legitimacy of their national laws. This process is even evident within the EU, where the Member States explicitly inserted into the TFEU a reservation that will ensure their ability to control their own energy mix and support systems going forward.

That said, over time the theory suggests that the regulatory support mechanisms should become more similar as countries seek to engage in either a 'race to the bottom' or a 'race to the top'. Despite this, there has never been an

[132] For example, a Senior Executive from a large wind turbine manufacturer has previously complained to the author about informal local content clauses and the preferential treatment of domestic firms in the bidding for Korean onshore and offshore windfarms. See also *Reuters*, 'Foreign Firms Cry Foul Over China Wind Power Rules', *Reuters* (online), 14 May 2009 <http://in.reuters.com/article/2009/05/14/china-wind-idINPEK31992120090514>.

[133] Aggarwal et al., above n 120, 233.

[134] Reuters, 'Foreign Firms Cry Foul over China Wind Power Rules', above n 132.

[135] Aggarwal et al., above n 120, 233–4.

attempt to unify the renewable energy laws of two countries and two, possibly three, attempts to harmonise regulatory support mechanisms at the EU level have failed. Instead, the EU has decided to try to foster a coordinated and cooperative approach between Member States in order to encourage a soft convergence of their support schemes over time. The move by Sweden and Norway to establish a Joint Support Scheme for their electricity certificates is an example of how this soft convergence may occur. Further research is required to understand more fully the processes of cooperation and coordination between other countries, especially given the establishment of the IRENA, which facilitates this process.

The other phenomenon that has recently come to the fore is that of explicit and very intentional divergence through the process of regulatory competition. This process is most evident among the technology innovators. This often reflects national industrial policy goals such as bolstering the domestic industry through the imposition of local content clauses or the provision of preferential subsidies to nationals of the country. The huge growth of the Chinese wind turbine manufacturing industry under both preferential subsidies and local content requirements, and the loss of market share by foreign firms, suggests that adopting divergent regulatory support mechanisms can lead to substantial financial benefits for a country.

7

Conclusion

The renewable energy sector has experienced an unprecedented boom over the past fifteen years. This research has shown that the reasons for this include the role that renewable energy plays in ensuring security of supply, addressing climate change, sustainably meeting rising energy demand, promoting economic growth and industrial policy issues.

Renewable energy laws have played a key role in this growth by providing regulatory support mechanisms such as research and development funding, feed-in tariffs, competitive tendering, renewable portfolio standards and tax incentives to assist with the development, commercialisation and accelerated deployment of new forms of renewable energy generation. This research is the first comparative analysis of the national framework legislation governing or promoting renewable electricity generation in all applicable countries.

The research question posed at the beginning of the book was whether, in light of the growing convergence of technologies adopted in the renewable energy sector, the laws in the sector would also show significant similarities. In order to answer this question, this book mapped the current position of national renewable energy laws in each of the 113 countries that have adopted such laws, as well as the relevant EU Directive and the IRENA Statute. This analysis was used to develop an understanding of the range of positions that exist for how countries legislatively define renewable energy, what they are trying to achieve through the adoption of these laws, and which regulatory support mechanisms they use. Based on this understanding of the different conceptions of renewable energy and drivers for countries legislating in the sector, an assessment was then made as to whether national renewable energy laws were likely to actively compete through regulatory competition, diverge, converge through cooperation and coordination processes or harmonise as the renewable energy technologies continue to converge around the world.

The initial hypothesis advanced at the opening of this book was that, as different techniques for generating renewable energy became commercialised and the manufacturing of renewable technologies became more concentrated in particular countries, renewable energy laws would also come under pressure to harmonise to facilitate trade, improve information sharing and ease administration. The initial hypothesis that there would be pressure for countries to harmonise their laws has proven correct, at least initially in the EU, although the EU Member States have been reasonably successful in resisting this pressure. This has meant that the EU has had, at least temporarily, to drop its preference for harmonisation in favour of adopting cooperation and coordination. Though there is evidence to suggest that the European Commission may be using the State Aid Guidelines to try and achieve a similar outcome. Meanwhile, there do not appear to have been any concerted efforts to harmonise the national renewable energy laws of any countries outside of the EU. Rather, some countries around the world have proactively pursued a concerted policy of regulatory competition. Indeed, contrary to the initial hypothesis, this research demonstrates that some significant national differences still exist within national framework pieces of legislation governing the promotion or accelerated deployment of renewable energy.

Throughout this book it has been argued that these national differences may largely be explained on the basis that many countries use their renewable energy laws to address national problems: in particular, those associated with energy security, dependence on fossil fuels and nuclear imports and the need to diversify their supply. Other examples of national problems that countries are seeking to address through these laws include encouraging economic growth, supporting sustainable development, ensuring system safety and reliability, as well as pursuing industrial policy objectives. As a result, it is not surprising that national laws have differing priorities and have developed divergently, especially among those countries that prioritise energy security. These divergent approaches have been demonstrated in the research in the different legislative objectives chosen,[1] the different priorities for the legislative objectives[2] and the different combinations of regulatory support mechanisms adopted.[3] One other notable difference is that some of the countries – most notably the technology innovators such as China – actively engage in regulatory competition through their renewable energy laws in order to further their own industrial policy objectives. Combined, these differences mean that

[1] See Chapter 4.
[2] See Chapter 4.
[3] See Chapter 5.

there is unlikely to be broad international convergence or harmonisation of national renewable energy laws in the short to medium term.

The EU Member States provide the exception to this rule. This may be explained by the fact that under the guise of the EU these countries have adopted a regional approach to both energy security through the development of the internal energy market, as well as climate change through the operation of the EU ETS. As a result of this regional cooperation and their coordinated approach to renewable energy, these countries have adopted similar legislative definitions, have similar legislative objectives, as well as national renewable energy targets and regular monitoring at and reporting to the EU level. Despite this, the EU countries have resisted giving the EU an unfettered 'shared competence' in the field of energy, with a reservation in Article 194(2) of the TFEU enabling Member States to 'determine the conditions for exploiting its energy resources, its choice between different energy sources and the general structure of its energy supply'.[4] This explains why Member States continue to have different designs and combinations of regulatory support mechanisms. However, even with these different regulatory support mechanisms, due to the imposition of the State Aid Guidelines, the presence of common goals and a regional approach to dealing with the externalities associated with renewable energy, some soft convergence has been evident.[5] This is particularly noticeable in the design of competitive tenders and feed-in

4 *Treaty on the Functioning of the European Union*, opened for signature 7 February 1992, [2009] OJ C 115/13 (entered into force 1 November 1993) *('TFEU')* Art 194(2).

5 See e.g. Lena Kitzing, Catherine Mitchell and Poul Erik Morthorst, 'Renewable Energy Policies in Europe: Converging or Diverging?' (2012) 51 *Energy Policy* 192; Gustav Resch et al., 'Coordination or Harmonisation? Feasible Pathways for a European Res Strategy Beyond 2020' (2013) 24 *Energy and Environment* 147; David Jacobs, *Renewable Energy Policy Convergence in the EU: The Evolution of Feed-in Tariffs in Germany, Spain and France* (Ashgate Publishing, 2012); Pablo del Río et al., 'Key Policy Approaches for a Harmonisation of RES(-E) Support in Europe – Main Options and Design Elements' (Report, European IEE Project Beyond2020, March 2012); Sian Crampsie, 'Renewables Convergence?' (2011) 34(14) *Utility Week* 9; Tatiana Romanova, 'Legal Approximation in Energy: A New Approach for the European Union and Russia' in Caroline Zuzemko, Andrei V Belyi, Andreas Goldthau and Michael F Keating (eds.), *Dynamics of Energy Governance in Europe and Russia* (Palgrave Macmillan, 2012) 23; Miquel Muñoz, Volker Oschmann and J David Tàbara, 'Harmonization of Renewable Electricity Feed-in Laws in the European Union' (2007) 35 *Energy Policy* 3104; Roger Hildingsson, Johannes Stripple and Andrew Jordan, 'Governing Renewable Energy in the EU: Confronting a Governance Dilemma' (2012) 11 *European Political Science* 18; Malgorzata Alicja Czeberkus, *Renewable Energy Sources: EU Policy and Law in Light of Integration* (LLM Thesis, University of Iceland, 2013); Peter-Olof Busch and Helge Jörgens, 'Europeanization Through Diffusion? Renewable Energy Policies and Alternative Sources for European Convergence' in Francesc Morata and Israel Solorio Sandoval (eds.), *European Energy Policy* (Edward Elgar, 2012) 66.

tariffs within Europe,[6] as well as the development of the Joint Swedish-Norwegian Electricity Certificate Market.

7.1 WHAT IS RENEWABLE ENERGY? A CASE OF CONCEPTUAL CONSENSUS

The first part of the book sought to discover whether a common understanding of the concept of 'renewable energy' had developed in the laws of countries seeking to accelerate its deployment. The term 'renewable' is commonly understood to mean that the natural resource depletes at a rate equal to or slower than its replacement rate. As Chapter 2 highlighted, the most commercialised and widely adopted technologies were often captured by the legislative definitions of 'renewable energy'. Wind power, photovoltaic solar, concentrated solar thermal energy, biomass, landfill gas, sewage treatment gas and biogas and small-scale hydropower were all captured within the definitions of more than 70 per cent of countries with national renewable energy legislation. This represents a significant degree of consensus around the energy sources identified as 'renewable' within the legislation of the different countries.

This seemingly harmonious picture belies national differences for some renewable energy sources. For example, among those countries where small-scale hydropower was captured within their definition, variance existed as to what constituted the cut-off point for determining whether a hydropower project would be small-scale or large-scale hydropower, with the cut-off point varying considerably: ranging from a low of 3MW[7] to a high of 30MW.[8] Equally, some countries such as Bangladesh and Belarus explicitly incorporate firewood of any origin into their definition of biomass, whereas other countries such as Australia only include wood waste that is not derived from old growth forests. The differences were even more prominent when it came to other technologies used to generate renewable energy. This research found that many of the decisions about whether a source of energy would be considered to be renewable or not were not conducted on strict scientific bases or in accordance with consistent legal principles. Two reasons were identified in the research to account for these differences. First, in many instances, the definition of renewable energy adopted or advocated for by different countries reflected their indigenous renewable (and sometimes non-renewable) energy

[6] See e.g. Jacobs, above n 5; Muñoz et al., above n 5; Kitzing, above n 5.
[7] This is approximately the power generated by a single wind turbine.
[8] This would generate enough power for approximately 22,000 homes.

sources. For example, most landlocked countries, though not all, did not capture the ocean energies of wave, tidal, current and hydrothermal energies within their legislative definition. While this approach makes sense, and is in fact a predictable outcome, the same cannot be said when the energy sources involved are arguably not renewable. For example, the Swedish legislation acknowledges that peat is not a renewable energy source, with it falling outside of their legislative definition. However, on closer inspection of the Swedish legislation, there is a supplementary definition for 'renewable electricity', which is defined as 'electricity produced from renewable energy sources or peat'.[9]

The second reason for these differences is that many countries seem to conflate a requirement into their definition of renewable energy that the energy must also be environmentally sustainable, and/or highly commercialised. The full extent of this conflation is not indicated by the small number of countries that have a definition that refers to environmentally sustainable development. Rather, this conflation must often be inferred from the presence or absence of qualifying renewable energy sources. For example, nearly a third of countries with renewable energy laws were found not to capture large-scale hydropower within their definition of 'renewable energy', which is often explained on the basis of environmental and social concerns. These concerns range from the loss of biodiversity, the impact on local groundwater flows, the carbon dioxide emissions from decaying plant life in the dams, the negative impact on water quality, the greater risk of natural disasters in the areas surrounding hydropower dams, the social impact of displacing people to build the dams and the natural constraints on the growth of large-scale hydropower. Meanwhile, ocean and riverine energies, which arguably are environmentally sustainable, were not accepted by many developing countries, presumably because of the costs involved with supporting emerging and non-commercialised technologies. The debates as to whether geothermal energy and nuclear energy, both sources of energy that deplete at such a slow rate as to be essentially inexhaustible, provide other examples of the conflation of the notion of sustainability into the definition of renewable energy. Despite geothermal and nuclear energy having similar depletion and replacement rates, ninety-six countries capture geothermal energy within their definition, while only Ecuador captures nuclear energy in its definition, largely due to public concern about safety and the environmental impacts of storing nuclear waste. This largely replicates the higher level of support for geothermal hot springs or aquifers, but not every country that supports geothermal energy has

[9] *Lag om elcertifikat* [Electricity Certificates Act] (Sweden) No 2011:1200, s 2 [1–2].

these forms of energy. It will be interesting to see whether the levels of public support change in the countries that will have to rely on hot dry rock technology, which is still not highly commercialised and often requires fracking, as the public become more aware of the actual processes used to generate energy and their environmental risks.

These outcomes, especially when arguably non-renewable sources of energy are incorporated into the legislative definitions of renewable energy, lead to illogical results that create a distance between the legislative definition and the commonly understood meaning of the term among the general population. However, despite these examples of divergence in the legislative definitions, the fact remains that there is an overwhelming level of consensus about the most commercialised forms of renewable energy being recognised as renewable energy sources.

Arguably, one of the most important prerequisites for convergence is that countries are seeking to legislate on the same subject matter. As noted by Chirico and Larouche, conceptual divergence is not always the result of a deliberate choice and may lurk beneath the surface and not be immediately perceptible.[10] This research suggests that there is a degree of international consensus about what energy sources are renewable for the purposes of electricity generation. This conceptual consensus as to the subject matter of the laws may provide the basis for later international harmonisation or legislative convergence in years to come. This suggests that there is an opportunity for greater efficiencies to be achieved within the renewable energy sector through the movement to a standard approach to the form and content of the legislative definition of renewable energy internationally. If this occurs, it will help to facilitate international trade, information sharing and administration, as well as the development of best practice within the international renewable energy industry.

7.2 WHY INTERVENE IN THE RENEWABLE ENERGY SECTOR? A CASE OF NORMATIVE DIVERGENCE

The second part of the book examined the justifications derived from economic theory for regulatory intervention into the renewable energy sector and then compared this to the legislative objectives contained in the renewable energy laws of countries who have legislated in this area. In this part, the

[10] Filomena Chirico and Pierre Larouche, 'Convergence and Divergence, in Law and Economics and Comparative Law' in Pierre Larouche and Peter Cserne (eds.), *National Legal Systems and Globalization* (Asser Press, 2013) 9, 10.

research was focused on normative convergence or divergence, i.e. whether countries intervene in the renewable energy sector for the same or differing reasons.

Chapter 3 examined the economic rationale for intervening in the renewable energy sector. Economic theory operates on the assumption that markets are the most efficient tool for allocating resources. As such, many economists believe that regulatory intervention into a market is only ever justified when it is required to address the sources of market failure. In the case of the renewable energy sector, economists have identified a number of market failures, as well as the presence of market barriers. The market failures in the sector are primarily due to the properties of electricity as a mixed good, that is, possessing characteristics of both a public and private good. There seems to be agreement that at least three market failures that pertain to the electricity sector and thus have an impact on the renewable energy sector exist to varying degrees in most countries around the world. First, there is the failure to price the negative externalities associated with fossil fuels such as the health and environmental impacts into the fossil fuel price, and conversely the failure to price the positive externalities associated with renewable energy such as its role in mitigating climate change and ensuring energy security into the renewable energy price. Second, the effects of positive spillovers and learning effects mean that private firms engaging in research and technological innovation in the sector may not receive the full return on their investments due to some of those benefits being publicly shared (or capable of being reverse engineered). The third market failure is due to information asymmetries resulting from hidden characteristics, principally in the form of uncertainty around future market developments and future generation costs for both fossil fuels and renewable energy.

Differences of opinion exist about the appropriate course of action to deal with market failures. Some economists believe that the risks of government failure, regulatory capture and imperfect information mean that regulatory intervention by the government into the sector may lead to suboptimal economic outcomes. Despite this, it appears that many economists and a majority of national governments around the world believe that some form of intervention, whether by legislation or policy, into the renewable energy sector is warranted. So how should that intervention be structured? From a purely economic perspective, approaches such as Pigovian taxes, which seek to address externalities by internalising the external costs of the activity not currently represented in the price paid by consumers, or the Coase Theorem, which argues that the most efficient outcome will be achieved when individuals are encouraged to negotiate outcomes, which eliminate externalities, are often preferred.

This research contends that, in reality, neither Pigovian taxes nor the Coase Theorem provides a perfect solution for addressing the market failures in the renewable energy sector. First, while some externalities associated with conventional fossil fuel electricity generation such as air pollution and carbon emissions may be taxed through Pigovian taxes such as emission trading schemes devised in accordance with 'polluter pays principles', not all externalities are easy to price. Indeed, Pigovian taxes are generally not seen as a solution for the externalities associated with energy security. Energy security, which is commonly achieved through diversifying supply and reducing dependence on imports of fossil fuels and nuclear energy, is one of the most challenging externalities to price. The value ascribed to energy security is normally a national political decision and the circumstances behind the decision are prone to change rapidly, in line with world events. Second, this research argues that the Coase Theorem relies upon unrealistic assumptions of perfect competition, no transaction costs and that both consumers and producers of externalities will be willing to voluntarily negotiate agreements leading to a socially optimal resource allocation. As a result, this research endorses the broad consensus that the most appropriate response to address the market failures associated with the renewable energy sector is regulatory intervention into the market.

Based on this, economic theory would seem to suggest that if countries were all legislating to address the same three market failures, then convergence, or even harmonisation, would be likely within the sector. This issue was tested in Chapter 4, which examined what countries were seeking to achieve by enacting national framework pieces of renewable energy law promoting the accelerated deployment of renewable energy by analysing the legislative objectives.

The research found that many countries were trying to achieve much more through their renewable energy laws than merely addressing the market failures identified by the economists. It also found that – contrary to the previous research conducted by Verbruggen and Lauber,[11] Haas et al.,[12] Mathews and Reinert,[13] Gallagher,[14] Lipp,[15] Aguirre and

[11] Aviel Verbruggen and Volkmar Lauber, 'Assessing the Performance of Renewable Electricity Support Instruments' (2012) 45 *Energy Policy* 635.

[12] Reinhard Haas et al., 'How to Promote Renewable Energy Systems Successfully and Effectively' (2004) 32 *Energy Policy* 833.

[13] John A Mathews and Erik S Reinert, 'Renewables, Manufacturing and Green Growth: Energy Strategies Based on Capturing Increasing Returns' (2014) 61 *Futures* 13.

[14] Kelly Sims Gallagher, 'Why & How Governments Support Renewable Energy' (2013) 142 *Daedalus* 59.

[15] Judith Lipp, 'Lessons for Effective Renewable Electricity Policy from Denmark, Germany and the United Kingdom' (2007) 35 *Energy Policy* 5481.

Ibijunle[16] and Charnovitz and Fischer[17] – countries seek to accelerate the deployment of renewable energy for a far broader range of reasons than has been identified previously. Further, countries also placed different weightings on their legislative objectives than the previous research had suggested.

Through studying the legislative objectives of the 113 countries with renewable energy laws, 28 separate categories of legislative objective were identified. There was significant national variance shown in the legislative objectives selected by different countries in their renewable energy laws. For example, sixteen different categories of legislative objective were identified among the five countries with such laws whose name starts with 'G'. This diversity represented the different problems that countries were trying to address through their legislation, as well as their level of economic development, and the level of environmental awareness of their citizens. Once the categories of legislative objective were identified, they were then categorised by their primary theme. Eight key themes of legislative objective were identified: security; sectoral; the economy; research, education and training; international and regional; the environment; industrial policy; and society. The frequency of each legislative objective and theme, and the relative weighting of the priorities assigned to them by each country, were then analysed.

The results of this research produced some unexpected results. For example, the fact that a theme of the legislative objectives sought to address one of the market failures identified within the energy sector did not mean that it was necessarily prioritised over those themes that merely sought to address domestic market barriers. The exceptions to this rule were the security themed objectives, which were often prioritised over the objectives in the other themes. However, the education, training and research themed objectives which target information asymmetries and positive spillovers were cited less frequently and were given a lower legislative priority than any of the sectoral themed objectives. This process was even evident in the environmental themed objectives. Indeed, despite the objective that most closely addressed the unpriced environmental externalities within the energy sector, 'reduce greenhouse gas emissions and address climate change', being frequently cited (twenty-eight countries) it was given a low priority, with a relative weighted rank of 5.89.

16 Mariana Aguirre and Gbenga Ibijunle, 'Determinants of Renewable Energy Growth: A Global Sample Analysis' (2014) 69 *Energy Policy* 374, 380.

17 Steve Charnovitz and Carolyn Fischer, 'Canada – Renewable Energy: Implications for WTO Law on Green and Not-so-Green Subsidies' (2015) 14 *World Trade Review* 177, 185.

A further unexpected result was evident when the relative frequency of citing each category of legislative objective and their relative weighted priorities were compared. For example, more countries addressed the issue of environmental protection in their legislative objectives (fifty-five countries) than energy security (forty-nine countries). However, the research also showed that the objectives addressing environmental protection were on average assigned a significantly lower priority, with a relative weighted rank of 5.09, compared to objectives addressing energy security, which had a weighted rank of 3.41. Thus while more countries had a legislative objective addressing environmental protection, it was commonly tacked on to the end of the objectives section as a form of political compromise.

In the context of each category of legislative objective, it became apparent that countries were largely legislating in their own self-interest, with their objectives strongly reflecting domestic concerns. For example, many of the countries that were seeking to address energy security issues were energy importers with low levels of energy self-sufficiency. However, some energy importers were also concerned with energy security, with a focus on maintaining 'security of demand' for the energy exports due to the important contribution it plays in terms of their government revenues. Another obvious domestic problem that was being addressed through these laws was an issue around concentrated supply from politically volatile regions. Countries that included the legislative objective of 'diversify supply' typically had very heavy dependence on energy supplies from a single country or region. In Europe, this meant that countries such as Finland were seeking to address the problem of Russia supplying 100 per cent of its gas and 88 per cent of its oil, while Japan was concerned about its dependence on the OPEC producers in the Middle East that supply 85 per cent of its oil. The research found that there was also a correlation between the timing of the legislation targeting diversity of supply, with many of them being introduced either during or in the immediate aftermath of the oil shocks of 2008.

Another feature of the legislative objectives contained in the national renewable energy laws was the sheer volume of legislative objectives that many countries were trying to achieve. On average countries have more than five categories of legislative objective in their national renewable energy law, however one country had more than fifteen categories of legislative objectives. This prompted two questions. First, are countries that are seeking to achieve multiple and conflicting legislative objectives far beyond the narrow economic justification engaging in over-regulation of the sector? Or is this phenomenon explained by politicians trying to keep diverse constituencies satisfied through including legislative objectives in a mere 'window-dressing' exercise? Second, does this provide an explanation for why countries have thus far been unwilling to engage in legal harmonisation or convergence of their renewable energy laws despite the globalisation of renewable energy technologies?

Once again, the countries with the greatest amount of similarity in their legislative objectives were the EU Member States. This again reflects a degree of normative convergence (i.e. alignment of cultural norms) that is derived from a regional approach to the accelerated deployment of renewable energy. Normative convergence or divergence around legislative objectives is important, as regulation seeks to achieve behavioural change. Cheng has argued that 'normative convergence represents the deepest kind of convergence and hence is the most elusive'.[18] As a result, normative convergence is critical to the success of convergence projects. For all other countries that do not have a degree of normative convergence, the differences in the legislative objectives, when coupled with the differing legislative definitions of renewable energy adopted in these laws, may have a profound impact on the scope, implementation and outcomes of these laws.

7.3 WHAT ROLE DO REGULATORY SUPPORT MECHANISMS PLAY IN NATIONAL RENEWABLE ENERGY LAWS? A CASE OF SUBSTANTIVE DIVERGENCE

In addition to the differences noted in the legislative definitions and legislative objectives, the forms and combinations of regulatory support mechanisms also varied significantly by country. This was discussed in Chapter 5. Many countries chose to structure their laws to include regulatory support mechanisms that would address the market failures, their specific market barriers and legislative objectives so as to level the playing field with conventional fossil fuels and to create a favourable environment for accelerated deployment. The forms of regulatory support mechanisms offered through renewable energy laws were found to include feed-in tariffs, feed-in premiums, renewable portfolio standards with tradeable green certificates, competitive tendering and auction bidding, net metering, subsidies, clean energy loans, rebates, tax incentives, public benefit funds, research and development support and green power schemes.

18 Thomas K Cheng, 'Convergence and Its Discontents: A Reconsideration of the Merits of Convergence of Global Competition Law' (2012) 12 *Chicago Journal of International Law* 433, 440.

When determining the correct regulatory support mechanisms to be adopted by a particular country, the research suggests that a number of factors be considered, including whether the regulatory support mechanism would:

- target the price or the quantity to be deployed;
- target the supply side or demand side of the market;
- be compulsory or voluntary;
- attempt to 'pick winners' or be technology neutral;
- be available industry-wide or targeted towards projects of a particular size or type;
- be capped by installation capacity, volume generated or a fixed budgetary pool;
- have its price fixed or be a market-orientated mechanism;
- be paid for by conventional utility companies, end-consumers or taxpayers more broadly;
- be subject to the rules of a regional organisation such as the European Union;
- be subject to the WTO Agreement on Subsidies and Countervailing Measures, the Energy Charter Treaty or other bilateral or multilateral investment treaty.

The vast majority of countries had adopted at least one primary regulatory support mechanism, generally competitive tendering, a feed-in tariff (a price-based mechanism), or a renewable portfolio standard (a quantity-based mechanism), with a number of secondary mechanisms. In some countries, hybrid mechanisms containing elements of different regulatory support mechanisms such as competitive tendering, FITs and tradeable green certificates have been developed.

Much of the previously reported research has focused on whether feed in tariffs/feed-in premiums are more efficient and effective than the use of a renewable portfolio standard, particularly in the context of the European Union. Initially, the European Commission held the view that renewable portfolio standards were a more effective tool as they would provide the renewable energy at least-cost. However, a number of studies have cast doubt on this claim, with feed-in tariffs proving to be able to deliver more installed generation at a lower cost. In response to significant resistance, the European Commission dropped its plans to harmonise the renewable energy laws of the Member States, instead preferring to encourage a form of 'soft convergence' through a process of cooperation and coordination. Until very recently, this led to feed-in tariffs being the preferred primary regulatory support mechanism in Europe. However, in 2014, the European

Commission drafted State Aid Guidelines that stipulated the countries should move to market based mechanisms such as feed-in premiums and competitive tendering. Since this time, there has been a shift towards competitive tendering among the EU Member States. This move to use competition law as a means of dictating which regulatory support schemes ought to be used has been labelled by some academics as another attempt at a 'top-down' harmonisation. This research suggests that the European Member States have shown that there are limits to harmonisation, especially when the Member States' national markets had organically adopted a range of differing approaches to intervening in the sector. As shown in Chapter 5, this extends to significant variance even when they do adopt the same regulatory support mechanism. In such cases, perhaps the more gradual approach of cooperation and coordination is more politically palatable to countries that feared external control over an essential element of their national economies.

Among the countries studied, high levels of divergence were evident in the regulatory support mechanisms adopted. This substantive divergence was found to be evident on two levels. First, in terms of the combinations of regulatory support mechanisms selected. While many countries seemed to follow the basic formula of one or two primary mechanisms and several secondary mechanisms, the combinations varied. Second, this divergence was evident in the structure of regulatory support mechanisms, which varied considerably by country. For example, while France provided a 30 per cent income tax deduction on the capital costs of renewable energy projects, Greece chose to stabilise the income tax coefficient instead of providing an income tax deduction.

Chapter 6 considered what all of this meant for the future development of regulatory support mechanisms contained in national renewable energy laws. It noted the considerable benefits, particularly for international market participants, of the regulatory support mechanisms within national renewable energy laws becoming more similar. This is because greater similarity in national renewable energy laws should lead to lower transaction costs, greater competition and, ultimately, to lower prices for the ultimate consumers. While the issue of whether regulatory support mechanisms are growing more similar or more divergent over time requires further research, from the case studies detailed above it is argued that some patterns are already evident. The starting position for most countries seems to be one of substantive divergence, with different regulatory support mechanisms being designed and implemented in different countries. This is explained as being a natural response to their different natural resources, legal traditions, governmental and socio-economic structures, and customs and norms, which in turn reinforces the legitimacy of their national laws. This process is even evident within the EU, where the Member States explicitly inserted into the TFEU a reservation which ensures their ability to control their own energy mix and support systems going forward.

That said, this research finds guarded support for the view of economic theorists that the regulatory support mechanisms will become more similar as countries seek to engage in either a 'race to the bottom' or a 'race to the top'. However, there has never been an attempt to unify the renewable energy laws of two countries and attempts to harmonise regulatory support mechanisms at the EU level have failed. Instead, the EU opted for fostering a coordinated and cooperative approach between Member States in order to encourage a soft convergence of their support schemes over time, with the recent move by Sweden and Norway to establish a Joint Support Scheme for their electricity certificates illustrating how such soft convergence may occur. However, the process is at best a slow or very gradual one, and given the very limited uptake of Joint Support Schemes there may not be convergence at the pace at which changes need to occur for an efficient global energy transition. Another source of knowledge that may lead to convergence around how countries design and modify their laws is likely to be the current investor–state disputes that have recently resulted from the Energy Charter Treaty and other bilateral and multilateral investment treaties. Further research is required to understand better whether these disputes are leading to regulatory changes, as well as the cooperation and coordination between other countries, especially given the establishment of the IRENA, which facilitates this process.

The other phenomenon identified from the research reported in this book is that a number of technology innovators appear to be intentionally encouraging divergence through the process of regulatory competition. This often reflects national industrial policy goals such as bolstering the domestic industry through the imposition of local content clauses or the provision of preferential subsidies to nationals of the country. The example of the substantial growth of the Chinese wind turbine manufacturing industry under both preferential subsidies and local content requirements, and the loss of market share by foreign firms, suggests that adopting divergent regulatory support mechanisms can lead to substantial financial benefits for a country. The relative weight that should be attached to the process of soft convergence, as compared to the rise of regulatory competition in regulatory support mechanisms, is therefore not yet clear. However, what is clear is that substantive divergence currently remains the base position for regulatory support mechanisms in national renewable energy laws and that further research is required to better understand this area.

7.4 CONCLUSION

This is the first research to examine the national renewable energy laws of all 113 countries that had such laws on 1 August 2018. It has challenged many of the existing understandings within the sector to provide a clear global picture of the emergent patterns, themes and tensions that have arisen from the national renewable energy laws. It has examined which countries legislated to promote renewable energy, what they considered to be renewable energy, why they considered that regulatory intervention in the renewable energy sector was warranted and how they intervened into the energy markets to support its accelerated deployment.

This research established that, despite the increasing convergence around the technologies and sources used to generate renewable electricity, there has not been the same degree of convergence in the national legislative approaches to governing or promoting the accelerated deployment of renewable energy. Countries have a much broader range of reasons for legislating than was previously understood; and they are shown to design, implement and combine their regulatory support mechanisms to meet their own domestic needs. As a result, normative convergence is unlikely to occur in the short to medium term as the vast majority of countries are dependent on energy imports and are concerned with bolstering their own domestic production rather than facilitating international trade. Many countries associate energy security, the highest weighted ranked legislative objective, very strongly with notions of national security and sovereignty. Thus they are unlikely to diminish their control over their energy sector unless it is necessary to secure supply. The regulatory competition that is evident among the technology innovators also perpetuates this process, as there is currently sufficiently aggressive competition in the market to mean that countries are not going to be willing to lose any advantage they may have by harmonising their laws. The Member States of the EU provide the exception to this rule, where a process of soft convergence through cooperation and coordination is evident. This may be explained by their shared cultural norms and values and regional pursuits of energy security, an internal energy market and environmental objectives, which are reflected in their similar legislative definitions and objectives.

Index

Australia
hydropower, 39
planning permission and approvals, 87–88
woody biomass, 29–32
Australian Greenhouse Office, 31
Australian Productivity Commission, 84

bioelectricity. *See also* biomass
producers, 28
biogas, 34–35
definition, legislative, 34
biomass, 27–34
advantages, 28
China, 27–34
definition, 27, 29
definition, legislative, 28, 34
direct combustion, 27
disadvantages, 28
generally, 28
health impacts, 34
liquefaction, 27
non-plantation native forest, 29–32
old growth forest, 29–32
regulatory issues, 30
sustainability criteria, 29
traditional, 32–34
woody, 29–32
biopower. *See* biomass
Brayton cycle, 26
Brazil
bioelectricity producer, 28
concentrated solar thermal technology, 26
hydropower, 39
photovoltaic solar, 24

carbon tax, 95
China
bioelectricity producer, 28
grid-connected projects versus installed capacity, 171
hydropower, 39, 41
regulatory competition, 16
wind farms, 171
climate change, 1
concentrated solar thermal technology. *See* solar energy
current energy, 49

distribution networks, 86

Ecuador
nuclear energy, 58–59
electricity. *See also* electricity sector; regulation; renewable energy sector
characteristic warranting regulatory treatment, 65–73
demand, 66, 71
economic development, 66–67
generation, interdependent with other fuel sources, 66, 68–69
geopolitical volatility, 68
individual welfare, 66–67
interdependent, 66
Large Combustion Plant Directive, 68
market concentration, 66, 72–73
national markets, 66
regulatory treatment, characteristics warranting, 65–73
stakeholders, 67
storage systems, 71–72
transmission and distribution networks, 71

electricity pricing
  externalities, social and environmental, 69
  inaccurate, 66
  information asymmetries, 66
electricity sector
  concentration, 72–73
  high barriers to entry, 72–73
  highly politicised, 67
  market barriers, 79–93
  market failures, 73–79
  storage systems, 71–72
  subsidies to fossil fuel, 80–82
  supply, global, 1
energy security, 1, 112–118, 264
  definition, 112
  diversify supply, 115
  exporters, 114
  fossil fuel imports, reduce the use of, 116
  geopolitical and economic factors, 113
  importers, 115
  indigenous energy sources, 118
energy storage systems, 71–72

feed-in tariffs, 167–223, *See also* regulatory support mechanisms
Finland
  concentrated solar thermal technology, 26
  peat, 55
  photovoltaic solar, 24
forest, native
  degradation, 29
  old growth, 29–32
  regulatory issues, 30
  sustainable logging, 32
fossil fuels
  costs, 3
  indigenous, 66
  nuclear generation, 80–82
  subsidies for, 80–82
fuel cells, 53–54, *See also* regulatory support mechanisms

geothermal energy, 43–47, *See also* regulatory support mechanisms
  aquifers, 44–45
  definition, legislative, 43
  environmental impacts, 46–47
  generally, 43
  hot fractured rock technology, 45–46
  hot springs, 44–45
  Iceland, 45
  impacts, 46–47
  Kenya, 45
  Philippines, 45
  risks, 46–47
Germany
  bioelectricity producer, 28
governance. *See* regulatory systems
green
  certificate trading, 13, 187–193
  power schemes, 13, 214–216

harmonious construction, 161
hydrogen fuel cells, 53–54
hydropower, 36–43, *See also* renewable energy sources
  advantages and disadvantages, 37
  Australia, 39
  Brazil, 39
  China, 39, 41
  classification of projects, 37, 42
  definition, legislative, 37, 38, 39, 42
  environmental impacts, 39–40
  generally, 36
  large scale, 38–39, 42, 61
  mature technology, 36
  pumped, 42–43
  small scale, 36–38
  social impacts, 40–41
  super profits, 41–42
  wildlife impact, 40
hydrothermal energy, 50–53
  definition, legislative, 53
  inconsistent use of terminology, 52

Iceland
  geothermal energy, 45
India
  bioelectricity producer, 28
  labour, lack of skilled, 91–92

Japan
  bioelectricity producer, 28
  planning permission and approvals, 87–88
  regulatory competition, 16

Kenya
  geothermal energy, 45
  traditional biomass, 33

landfill gas, 34–35, *See also* renewable energy sources
definition, legislative, 34
Large Combustion Plant Directive, 68
laws. *See* renewable energy laws
legal mechanisms. *See* regulatory support mechanisms
legislative objectives in renewable energy law, 250–253, 264, *See also* renewable energy laws
diversify supply, 112–118
economic, 122–132
education, training and research, 132–135
energy security, 112–118
environmental, 138–144
harmonious construction, 161
industrial policy, 145–152
international agreements, 135–138
regional integration, 135–138
research, 103–112
sectoral, 118–122
security, 112–118
social, 152–159

Malaysia
concentrated solar thermal technology, 26
wind energy, 20
market barriers, 79–93, *See also* renewable energy sector
addressing, 99
Australian Productivity Commission, 84
economies of scale, 89
fossil fuels, subsidies for, 80–82
impact on renewable generation, 85
labour, lack of skilled, 91–92
nuclear generation, subsidies for, 80–82
planning permission and approvals, 87–88
policy instrument, 83
policy uncertainty, 82–85
principal-agent problem, 90–91
regulatory uncertainty, 82–85
split incentives, 90–91
subsidies, 80–82
transmission and distribution networks, access to, 85–87
market failures
addressing, 94–96, 99
finance, limited access to, 89–90
risk, appropriately pricing, 89–90
mechanisms. *See* regulatory support mechanisms

Netherlands
planning permission and approvals, 87–88
nuclear energy, 56–60, *See also* renewable energy sources
Ecuador, 59
low-carbon energy source, 58
non-renewable, 57, 60
regulatory support, 56
subsidies for, 80–82
United Kingdom, 56

ocean currents. *See* tides, waves and ocean currents

Paris Agreement, 1
Nationally Determined Contributions, 1, 70
peat, 55, *See also* renewable energy sources
definition, legislative, 55
Finland, 55
Sweden, 55
Philippines
geothermal energy, 45
photovoltaic solar energy, 24–26
advantages, 24–25
Brazil, 24
disadvantages, 25–26
Finland, 24
planning permission and approvals. *See also* renewable energy sector
Australia, 87–88
Japan, 87–88
Netherlands, 87–88
wind projects, onshore, 88
power generation
global, 1

Rankine cycle, 26
regulation. *See also* electricity sector; regulatory systems; renewable energy laws; renewable energy sector
agencies, 85
economic justification, 65–97
economic perspective, warranted from, 94
electricity, characteristics that warrant, 65–73
market failures, 73–79
regulatory agencies, 85
regulatory support mechanisms, 3, 4, 8, 13–14
auction bidding, 193–198
capped, 168
classification, 13, 167–175

regulatory support mechanisms (cont.)
  clean energy loans, 205–207
  coase theorem, 94–96
  competitive tendering, 13, 193–198
  compulsory or voluntary, 168
  development, 14
  energy storage systems, 72
  evaluating, 219–222
  feed-in premiums, 13, 185–187
  feed-in tariffs, 13, 176–185
  future development, 224–230
  green certificate trading, 13
  green power schemes, 13, 214–216
  indirect, 13, 175
  industry-wide, 168
  investment tax credits, 13, 210
  investments versus operating support, 170
  multiple, 13
  net metering, 198–200
  pigovian taxes, 94, 99
  price driven, 167, 168, 173–175
  primary or secondary instrument, 170
  quantity driven, 167, 173–175
  quota system, 13, 187–193
  rebates, 207–208
  regulatory competition, 241–248
  renewable energy targets, 200–202
  renewable portfolio standards, 13, 187–193
  renewal energy credits, 13
  research and development support, 212–214
  role, 167–223
  selection, 167–175
  storage systems, 72
  subsidies, 202–204
  supply or demand, 167
  tax incentives, 208–211
  technology neutral, 168
  tradeable green certificates, 187–193
  types, 176–219
regulatory systems. *See also* renewable energy laws
  competition, 241–248, 264
  convergence, 237–240, 250–264
  divergence, 240–241, 255–260
  harmonisation, 231–237, 250–264
  regulatory competition, 264
  support mechanisms. *See* regulatory support mechanisms
  unification, 230
renewable energy. *See also* regulatory systems
  definition, 19–20, 29–32
  definition, legislative, 7–10, 19, 60–62
  national law, analysis of all countries, 7–10
  scientific meaning, 62
renewable energy laws, 7, 98–163, *See also* regulatory systems
  analysis, 10, 65, 250–264
  convergence, 237–240, 250–264
  definition, legislation, 7–10
  divergence, 240–241, 255–260
  economic objectives, 122–132
  education, training and research objectives, 132–135
  environmental objectives, 138–144
  harmonisation, 7–10, 231–237
  industrial policy objectives, 145–152
  international and regional objectives, 135–138
  legislative objectives, 98–163
  rationale for legislating, 98–163
  regulatory competition, 241–248
  regulatory support mechanisms. *See* regulatory support mechanisms
  resolving conflict between competing, 159–161
  sectoral objectives, 118–122
  security objectives, 112–118
  social objectives, 152–159, 264
  unification, 230
renewable energy sector, 1–7, *See also* electricity; electricity sector
  connect then manage, 87
  fossil fuel costs, 3
  government intervention, 4, 65, 94
  growth, 1–2, 65
  information asymmetries, 3–4, 74, 77–79
  investment, 1–3
  Large Combustion Plant Directive, 68
  market barriers, 79–93
  market failures, 3–4, 66, 73–79
  planning permission and approvals, 87–88
  principal-agent problem, 90–91
  reforms, 82
  regulatory models, 5
  regulatory support mechanisms, 3
  spillovers and learning effects, 3–4, 74–76, 82
  split incentives, 90–91
  subsidies to fossil fuel, 80–82
  technology and equipment costs, 3
  unpriced externalities, 3–4, 74, 76–77, 92, 99
renewable energy sources, 8, 19–62, *See also* energy security

biogas, 34–35
biomass, 27–34
commercialised, highly, 60–62
definition, 19–20, 29–32
environmental impacts, 19
fuel cells, 53–54
generally, 19–20
geothermal energy, 43–47
hydrogen fuel cells, 53–54
hydropower, 36–43
landfill gas, 34–35
non-renewable, 60–62
nuclear energy, 56–60
peat, 55
sewage treatment gas, 34–35
solar energy, 24–27
sustainable, 60–62
tides, waves and ocean currents, 47–53
wind energy, 20–23
riverine energy. *See* tides, waves and ocean currents

security. *See* energy security
sewage treatment gas, 34–35
definition, legislative, 34
solar energy
advantages, 27
Brazil, 26
categories, 26
definition, legislative, 26
Finland, 26
generally, 24, 26
Malaysia, 26
photovoltaic, 24–26
Stirling cycle, 26
Sweden
peat, 55

tax
incentives, 208–211
investment, 210
pigovian, 94, 99
tidal power, 49
tides, waves and ocean currents, 47–53, *See also* renewable energy sources
definition, legislative, 48, 49
hydrothermal energy, 50–53
maremotermica, 50–53
ocean thermal layering, 50–53
osmotic energy, 53
salt gradient, 53
tidal power, 49
wave energy, 49–50
transmission and distribution networks, 85–87

United Kingdom
bioelectricity producer, 28
nuclear energy, 56
Transmission Access Review, 87
transmission and distribution networks, 86

wave energy, 49–50, *See also* tides, waves and ocean currents
definition, legislative, 50
wet biomass conversion, 34, *See also* biomass
wind energy, 20–23
acceptance, lack of, 92–93
definition, legislative, 20
environmental impacts, 22–23
floating platforms, 21
generally, 20
health impacts, 23
innovations, 21
Malaysia, 20
noise emissions, 23
onshore and offshore, 21–22
planning permission and approvals, 88
process, 21
types, 21
unpriced externalities, 92–93
wind farm siting, 21–22
wind power and, 21–22
Wind Turbine Syndrome, 23
Wind Turbine Syndrome, 23

For EU product safety concerns, contact us at Calle de José Abascal, 56–1°, 28003 Madrid, Spain or eugpsr@cambridge.org.

www.ingramcontent.com/pod-product-compliance
Lightning Source LLC
LaVergne TN
LVHW011809200926
848484LV00005B/1115

* 9 7 8 1 1 0 7 1 8 5 7 6 0 *